TUTORIAL CHEMISTRY

4

d- and f-Block Chemistry

CHRIS J. JONES
University of Birmingham

ISBN 0-85404-637-2

A catalogue record for this book is available from the British Library

Reprinted in 2006

Publlished by The Royal Society of Chemistry, Thomas Graham House, Science Park,
Milton Road, Cambridge CB4 0WF, UK
Registered Charity No. 207890
For further information see our web site at www.rsc.org

Typeset in Great Britain by Wyvern 21, Bristol
Printed and bound by Henry Ling Ltd, Dorchester, Dorset, UK

Preface

At first sight it may seem strange that metallic elements such as iron or copper, which we encounter in machinery or electrical wiring, might also be intimately involved in the function of living organisms. However, our very survival depends upon the particular chemical properties of these metallic elements, which belong to the d-block series known as the 'transition metals'. To illustrate this point we might note that, after every breath we take, the oxygen we inhale is collected by the red protein haemoglobin present in our blood. The red color of blood arises from the presence of iron in this protein, and it is to this iron that the oxygen becomes attached. Through its attachment to hemoglobin, oxygen is transported in the blood to sites in the body where another protein containing both iron and copper effects its reduction. This releases energy as part of the process of respiration. What is so special about iron and copper that biology has selected them for these particular rôles? Why not nickel rather than iron? What do we need to know about their chemistry to understand why they are so specially suited to their particular functions?

Apart from their rather immediate importance in maintaining life, transition elements are of major importance in enhancing the quality of our daily lives. Among the materials we encounter each day, many either contain transition elements or have been formed in processes which involve them. Even plastic materials may have required the use of a transition metal catalyst in their manufacture.

In order to understand the biological rôles of transition metals, or develop new chemical processes involving them, it is first necessary to understand the principles which underly the chemistry of these elements. In a short text such as this it is not possible to describe the chemistry of the transition elements in any comprehensive way, and no attempt is made to do so. Rather, it is the aim of this book to introduce some basic principles which would allow a student of the subject to make more sense of the varied, and sometimes surprising, chemistry they may encounter when reading more comprehensive accounts of transition element chemistry. Several short primers already exist which provide an introduction to the d- and f-block elements and it is not the aim of this text to duplicate these, but rather to complement them.

The text is intended to take the reader through some of the topics covered in the first two years of an undergraduate course in chemistry. It assumes some basic knowledge of topics which should be covered in other courses at this level. These include atomic structure, simple quantum theory, simple thermodynamic relationships and electrode potentials. A knowledge of group theory is not explicitly required to follow the text, but reference is made to the symbols of group theory. The data in the text are based on published sources. However, it should not be assumed that data in the problems are based on actual measurements. Although reported data have been used where possible, some values have been calculated or invented for the purpose of the question.

As with all works of this type, the final text does not reflect the input of the author alone. My thanks go to Matt Barton, Andy Millar and Steve Vickers for checking a draft of the text, and to Martyn Berry for his helpful and insightful comments after reviewing the manuscript. However, the responsibility for any errors which remain is mine alone, although I assume no responsibility for the cover design which was beyond my control!

Finally, my thanks must also go to my long-suffering wife, Judy, for her forbearance and patience whilst my attentions have been redirected to yet another 'project'.

Chris J. Jones
Birmingham

TUTORIAL CHEMISTRY TEXTS

This series of books consists of short, single-topic or modular texts, concentrating on the fundamental areas of chemistry taught in undergraduate science courses. Each book provides a concise account of the basic principles underlying a given subject, embodying an independent-learning philosophy and including worked examples. The one topic, one book approach ensures that the series is adaptable to chemistry courses across a variety of institutions.

TITLES IN THE SERIES

Stereochemistry *D G Morris*
Reactions and Characterization of Solids *S E Dann*
Main Group Chemistry *W Henderson*
d- and f-Block Chemistry *C J Jones*
Structure and Bonding *J Barrett*
Functional Group Chemistry *J R Hanson*

Further information about this series is available at www.chemsoc.org/tct
Orders and enquiries should be sent to:
Sales and Customer Care, Royal Society of Chemistry, Thomas Graham House, Science Park, Milton Road, Cambridge CB4 0WF, UK
Tel: +44 1223 432360; Fax: +44 1223 426017; Email: sales@rsc.org

Contents

1
Introduction

Aims

By the end of this chapter you should understand the meaning of the term transition elements and have gained an appreciation of:

- The origins and terrestrial occurrence of the transition elements
- The historical development of transition element chemistry
- The importance and applications of the transition elements

1.1 What are the Transition Elements?

Within the Periodic Table (Table 1.1) the chemical elements can be grouped together in blocks according to occupancy of their outermost, or valence shell, atomic orbitals. Thus hydrogen and the alkali metals, lithium to francium, with a half-filled outer s subshell, together with helium and the alkaline earth metals, beryllium to radium, with a filled outer s subshell, comprise the s-block elements (Figure 1.1). Similarly, elements with a partly or fully filled outer p subshell comprise the p-block elements. That is the block from boron through to neon down to thallium through to radon. Together the s- and p-block elements comprise the main group elements.

Between these two blocks of elements there are two further blocks containing the transition elements. Strictly speaking, the term transition element applies to an element with a partly filled d or f subshell and so excludes those with d^0 or d^{10} and f^0 or f^{14} electron configurations. However, it is convenient to include copper, silver and gold in this classification as these elements commonly form ions with partly filled d subshells. Although their neutral atoms have d^{10} electron configurations, it is the chemistry of their ions which is of primary interest here. Similar arguments apply to ytterbium and nobelium. Their atoms have $f^{14}s^2$

Table 1.1 The Periodic Table emphasizing the transition series d-block and inner transition series f-block elements[a]

USA[b]	IA	IIA	IIIB	IVB	VB	VIB	VIIB	VIIIB			IB	IIB	IIIA	IVA	VA	VIA	VIIA	VIIIA
EU[c]	IA	IIA	IIIA	IVA	VA	VIA	VIIA	VIII			IB	IIB	IIIB	IVB	VB	VIB	VIIB	0
IUPAC[d]	1	2	3	4	5	6	7	8	9	10	11	12	13	14	15	16	17	18
	H	He																
	Li	Be											B	C	N	O	F	Ne
	Na	Mg											Al	Si	P	S	Cl	Ar
4s 3d	K	Ca	Sc	Ti	V	Cr	Mn	Fe	Co	Ni	Cu	Zn	Ga	Ge	As	Se	Br	Kr
4p			21	22	23	24	25	26	27	28	29	30						
5s 4d	Rb	Sr	Y	Zr	Nb	Mo	Tc	Ru	Rh	Pd	Ag	Cd	In	Sn	Sb	Te	I	Xe
5p			39	40	41	42	43	44	45	46	47	48						
6s 5d 4f	Cs	Ba	La	Hf	Ta	W	Re	Os	Ir	Pt	Au	Hg	Tl	Pb	Bi	Po	As	Rn
6p			57	72	73	74	75	76	77	78	79	80						
7s 6d 5f	Fr	Ra	Ac	Rf	Db	Sg	Bh	Hs	Mt									
7p			89	104	105	106	107	108	109	110	111	112						
				(Unq)	(Unp)	(Unh)	(Uns)	(Uno)	(Une)	(Uun)	(Uuu)	(Uub)						

Ce	Pr	Nd	Pm	Sm	Eu	Gd	Tb	Dy	Ho	Er	Tm	Yb	Lu
58	59	60	61	62	63	64	65	66	67	68	69	70	71
Th	Pa	U	Np	Pu	Am	Cm	Bk	Cf	Es	Fm	Md	No	Lr
90	91	92	93	94	95	96	97	98	99	100	101	102	103

[a]Atomic numbers *Z* are shown beneath the element symbol for the d- and f- block elements; International Union of Pure and Applied Chemistry (IUPAC) symbols are shown in parentheses for the fourth row of the d-block.
[b]This numbering system for groups has traditionally been used in the USA.
[c]This numbering system for groups has traditionally been used in Europe.
[d]This numbering system for groups has been aproved by IUPAC and is the one used throughout this text.

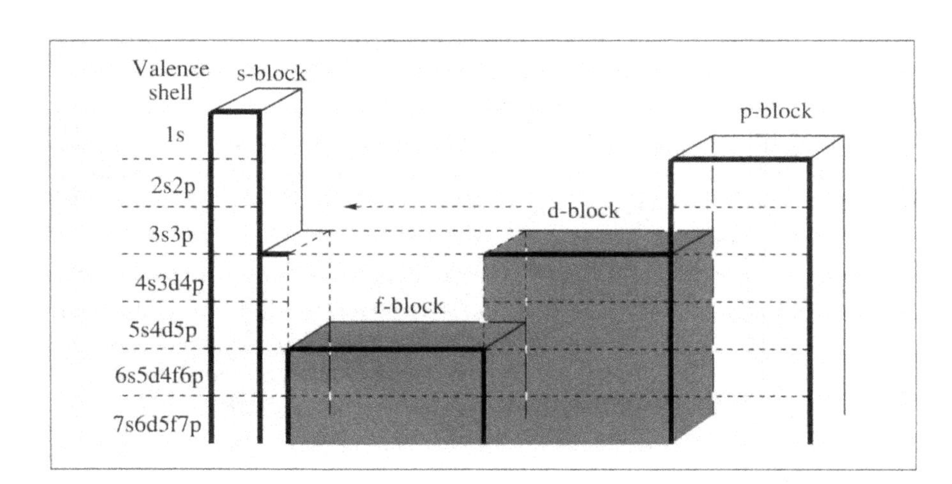

Figure 1.1 Blocks within the Periodic Table

valence shell electron configurations. However, it makes little sense to exclude them from the transition element series since both form stable trications with f^{13} electron configurations. Despite the fact that their atoms, and their stable dications, have d^{10} electron configurations, zinc, cadmium and mercury are often considered together with the transition elements. In effect they constitute the last members of the d-block series. The elements within the d-block are generally referred to as the transition metals. The elements from scandium to copper are often refered to as the first transition series or first-row transition metals, those from yttrium to silver form the second transition series or second-row and those from lanthanum to gold form the third series or row within the d-block.

The term **transition metal** is commonly used to refer to the d-block transition elements, Sc to Cu, Y to Ag and La to Au, but excluding the f-block metals. Here the term **transition element** will be taken to have the more general meaning and to include both d- and f-block elements.

The f-block elements comprise two series of inner transition elements which appear, firstly after lanthanum and secondly after actinium, in the Periodic Table. The elements from cerium to lutetium are known as the lanthanides and, because of its chemical similarity to these elements, lanthanum is usually included with them. Scandium and yttrium also show strong chemical similarities to the lanthanides, so that the chemistry of these elements is also often considered in conjunction with that of the lanthanide series. The second series of f-block elements, from thorium to lawrencium, is known as the actinide series and again it is usual to consider actinium together with this series.

In this text the symbol M will be used as a generic symbol denoting a d-block metal, Ln will be used as a generic symbol denoting a lanthanide element and An as a generic symbol to denote an actinide element.

The transition elements have several characteristic properties. All are metals which conduct heat and electricity well. All will form alloys with one another, and with metallic main group elements. Except for mercury, which is a liquid at room temperature, all appear as high melting point and high boiling point lustrous solids. Many of the transition elements react with mineral acids to form salts, though some of the d-block metals are rather inert in this respect (Box 1.1). Under the conditions accessible through simple laboratory procedures, most of the d-block elements can show more than one oxidation state. A rather different situation arises with the lanthanide elements, since their chemistry is dominated by Ln^{3+} ions. Like the d-block elements, the early actinides from protactinium to americium can show more than one oxidation state. However, the later actinides are more like the lanthanides in that their chemistry is dominated by An^{3+} ions. The presence of partly filled d or f subshells results in some transition element ions containing odd numbers of electrons, so that their compounds have magnetic properties. In addition, many form coloured compounds, the absorption of visible light being associated with the presence of the partly filled d or f subshell. The electron transfer, magnetic and optical properties of the transition elements are important features underlying their use in an amazing variety of applications.

Box 1.1 Dissolving Metals

During early studies of the transition elements it was found that some metals would not dissolve in concentrated nitric acid; these include ruthenium, osmium, rhodium, iridium, platinum and gold. However, it is possible to dissolve platinum or gold in *aqua regia*, a mixture of 1 part concentrated nitric and 3–4 parts concentrated hydrochloric acid. This gives the soluble compounds $H[AuCl_4]$ and $H_2[PtCl_6]$. Rhodium will dissolve in hot concentrated sulfuric acid, while both rhodium and iridium will dissolve in hot concentrated hydrochloric acid containing sodium chlorate. Ruthenium and osmium are inert to mineral acids below 100 °C but can be converted to the soluble salts $Na_2[RuO_4]$ and $Na_2[OsO_4(OH)_2]$, respectively, by fusion with an oxidizing alkali such as Na_2O_2. Among the earlier d-block elements, niobium and tantalum are also resistant to attack by mineral acids but will dissolve in mixtures of nitric and hydrofluoric acids; molybdenum and tungsten behave similarly. Zirconium and hafnium are fairly unreactive but will dissolve in hydrofluoric acid.

1.2 Where do Transition Elements Come From?

1.2.1 Origins

As with all of the naturally occurring elements, the transition elements originate from the nuclear reactions occurring within stars and supernovae. These reactions converted the hydrogen and helium, formed in the 'hot big bang', into the other chemical elements. Those elements having only isotopes with half lives substantially shorter than the time elapsed since their formation have been lost through radioactive decay. These include technetium, promethium and the actinides other than ^{235}U ($t_{1/2}$ = 7.0 × 10^8 y), ^{238}U ($t_{1/2}$ = 4.47 × 10^9 y) and ^{232}Th ($t_{1/2}$ = 1.41 × 10^{10} y), although traces of some elements may arise as a result of the radioactive decay of other longer lived isotopes, *e.g.* ^{227}Ac from ^{235}U.

Isotopes are nuclides which have the same number of protons, but different numbers of neutrons, in their nuclei. An isotope is defined by two numbers: the **mass number**, *A*, which is total number of nucleons (protons and neutrons) in the nucleus, and the **atomic number**, *Z*, which is total nunber of protons in the nucleus. The value of *A* is written as a superscript and of *Z* as a subscript preceeding the element symbol, *e.g.* $^{95}_{42}Mo$, $^{238}_{92}U$, $^{239}_{94}Pu$.

The development of cyclotrons and nuclear reactors in the middle of the 20th century made possible the production of radioactive isotopes which are not naturally present in any significant quantity on Earth. Thus in a nuclear reactor some of the neutrons released by uranium fission may be absorbed by ^{238}U, leading to the formation of ^{239}Pu. Similarly, the irradiation of molybdenum with neutrons gives ^{99}Mo which decays to the metastable γ-ray emitting nuclide ^{99m}Tc. This is of great importance in certain medical diagnosis applications. Some isotopes of the heavier actinide elements beyond plutonium can also be obtained from lighter actinides through successive neutron capture processes in high neutron flux nuclear reactors. In this way, gram quantities of californium and milligram quantities of einsteinium have been obtained. The elements beyond fermium are produced by bombarding heavy element targets with the nuclei of lighter elements which have been accelerated in a cyclotron.

Such methods have allowed all the elements of the 6d series, *i.e.* to $Z = 112$, to be produced. However, these heavy elements are not available in sufficient quantities for conventional chemical studies and so will not be considered further here. Beyond fermium, even the longest-lived isotopes of the elements are highly radioactive, having only short half-lives (*e.g.* 53 d for ^{258}Md, 185 s for ^{255}No and 45 s for ^{256}Lr).

1.2.2 The Terrestrial Abundances of The Transition Elements

The abundances in the Earth's crust of both the d-block transition metals and the f-block inner transition metals vary considerably, as shown in Table 1.2. Iron is the most common of the transition metals (6.30% by mass of the crustal rocks) and this reflects the high yield of iron from element synthesis reactions in stellar supernovae. Titanium (0.66%) and manganese (0.11%) are also quite abundant, but some of the heavier

Table 1.2 Estimated abundances of transition elements in the Earth's crust[a]

d-block									
Sc	Ti	V	Cr	Mn	Fe	Co	Ni	Cu	Zn
26	6600	190	140	1100	63000	30	90	68	79
Y	Zr	Nb	Mo	Tc	Ru	Rh	Pd	Ag	Cd
29	130	17	1.1	–	0.001	0.0007	0.0063	0.080	0.15
La	Hf	Ta	W	Re	Os	Ir	Pt	Au	Hg
34	3.3	1.7	1.1	0.0026	0.0018	0.0004	0.037	0.0031	0.067
f-block									
Ce	Pr	Nd	Sm	Eu	Gd	Tb	Dy	Ho	Er
60	8.7	33	6.0	1.8	5.2	0.94	6.2	1.2	3.0
Tm	Yb	Lu	Th	U					
0.45	2.8	0.56	6.0	1.8					

[a]Numerical values are in parts per million (ppm) from the WebElements Periodic Table of the Elements at the website created by Dr M. J. Winter: http://www.webelements.com/

transition metals, particularly ruthenium, osmium, rhodium, iridium and gold, are rather rare. In contrast, and despite their historic name of 'the rare earths', the lanthanide elements are not particularly rare. All except promethium are more abundant in crustal rocks than silver, gold or iodine, and most are of comparable abundance to bromine (3 ppm). The two primordial actinides, thorium and uranium, are of comparable abundance to the lanthanides, and considerably less rare than many of the heavier d-block elements.

Fortunately, for those wishing to extract transition elements from the Earth's crust, planet formation is not the only process leading to their segregation and concentration. The chemical and geochemical processes involved in forming the Earth's crust have also concentrated metals in ways which reflect their chemical properties. Some metals are rather unreactive under terrestrial conditions and may appear naturally in native form, as the elemental metal or alloyed with other elemental metals. Gold nuggets provide a particularly well-known example. Other metals form stable binary compounds with oxygen or sulfur and may be found in the form of oxides or sulfides. Iron pyrites, FeS_2, also known as 'fool's gold' because of the gold metallic lustre of its crystals, provides an example. Metal ions in oxidation state +3 tend to form insoluble phosphate salts and may be found in phosphate or vanadate minerals such as monazite, $LnPO_4$. Metal ions in oxidation state +4 may appear in silicates, exemplified by zircon, $ZrSiO_4$. These mineral deposits can provide commercially viable sources of the common metals such as vanadium, chromium, manganese, iron, nickel and copper. The rare metals ruthenium, osmium, rhodium, iridium, palladium and platinum, are also known as the 'platinum metals'. Together with rhenium, these are often present at low concentrations in the ores of more common metals. Consequently, they tend to be obtained as by-products during the processing of these ores and the purification of the metal produced.

Worked Problem 1.1

Q Outline the chemical steps of a process for separating the rare metals Ru, Os, Rh, Ir, Pd, Pt, Ag and Au present in the wastes ('anode slimes') formed during the electrolytic purification of nickel or copper.

A An answer to this question may be found by consulting Chapters 25, 26, 27 and 28 in the first edition of *Chemistry of the Elements*, by N. N. Greenwood and A. Earnshaw (1984). A possible separation process which exploits the differing reactivities and solubilities of the platinum group metals (Box 1.1) is shown in Scheme 1.1.

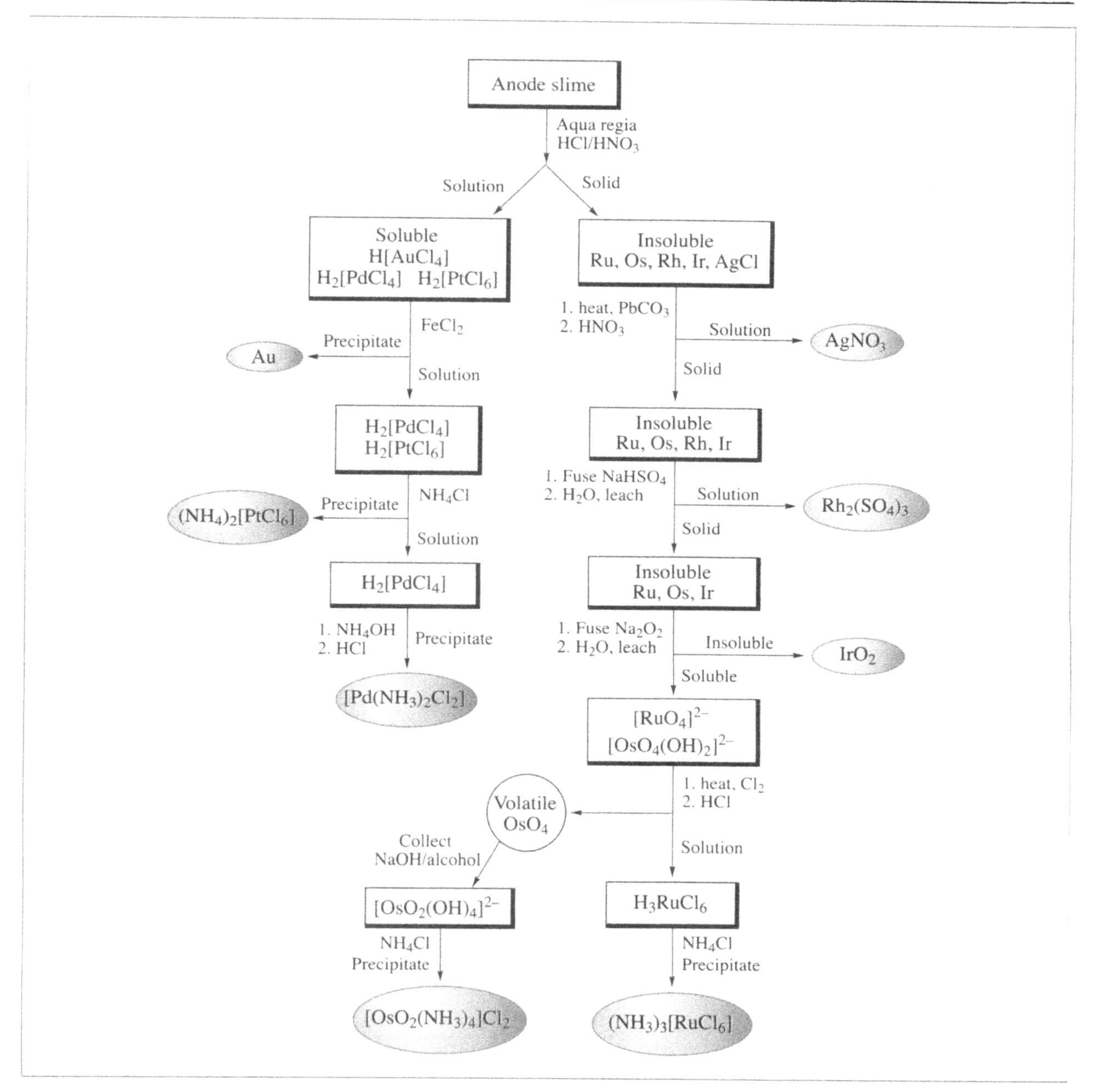

Scheme 1.1

1.3 The Historical Development of Transition Element Chemistry

Humans have been aware of the existence of some of the transition elements for thousands of years. Decorative beads made from iron meteorites date back 6000 years to 4000 BC and the use of copper almost certainly predates this. However, it is only in the 20th century that the complete set of stable elements has been identified and samples of pure metal isolated for each. The isolation of technetium, for example, was not achieved until 1937, within living memory for some. Such work continues to this day in studies of the formation and atomic properties of the 6d transition elements. The known or proposed dates of discovery of the transition elements are summarized in Table 1.3.

In broad terms, the discovery of the transition elements can be divided into four phases which reflect both the chemical nature of the elements and the development of human knowledge. The first metals to be discovered were those which can exist on Earth in native form, that is as the elemental metal. The iron beads mentioned above are not a typical example, in that iron is normally found on Earth in combination with other elements, particularly oxygen or sulfur. However, iron-rich meteorites provide an extraterrestrial source of impure metallic iron which was discovered at an early stage. The transition metals more typically found in native form are copper, silver and gold. These so-called 'coinage metals', being lustrous and malleable, would be conspicuous among other minerals to ancient peoples, gold in particular being prized for its lustre and resistance to corrosion. The use of copper probably dates back to *ca.* 5000 BC and it is thought that the Egyptians were using a form of gold coinage as early as *ca.* 3400 BC.

The second phase of transition element discovery involved those which could readily be released from minerals through heating or reduction by hot charcoal. Again copper in the carbonate mineral malachite, silver in the sulfide mineral argentite and mercury as the sulfide in cinnabar might

Table 1.3 Dates of discovery or first known use of the transition elements[a]

Sc	Ti	V	Cr	Mn	*Fe*	Co	Ni	*Cu*	*Zn*
1879	1791	1830	1797	1774	*1200 BC*	1735	1751	*5000 BC*	*13th C*
Y	Zr	Nb	Mo	**Tc**	Ru	Rh	Pd	*Ag*	Cd
1843	1789	1867	1778	**1937**	1844	1803	1803	*3000 BC*	1817
La	**Hf**	Ta	W	**Re**	Os	Ir	Pt	*Au*	*Hg*
1839	**1922**	1802	1781	**1925**	1803	1803	1736	*3400 BC*	*500 BC*
Ac									
1899									

Ce	Pr	Nd	**Pm**	Sm	**Eu**	Gd	Tb	Dy	Ho	Er	Tm	Yb	**Lu**
1839	1885	1885	**1947**	1879	**1901**	1880	1843	1886	1879	1843	1879	1878	**1907**
Th	**Pa**	U	**Np**	**Pu**	**Am**	**Cm**	**Bk**	**Cf**	**Es**	**Fm**	**Md**	**No**	**Lr**
1829	**1913**	1789	**1940**	**1940**	**1944**	**1944**	**1949**	**1950**	**1952**	**1952**	**1955**	**1958**	**1961**

[a]Elements known and used since before the 18th century are shown in *italic* text, those discovered in the 18th and 19th century in plain text and those discovered in the 20th century in **bold** text.

first have been obtained in metallic form from minerals mixed with the glowing embers of a camp fire. The formation of metallic copper through charcoal reduction appears to have been known by 3500 BC. The copper/tin alloy bronze is said to have been discovered before about 3000 BC, leading to the 'Bronze Age'. Brass, another copper alloy formed with zinc, appeared in Palestine around 1400 BC. Iron is superior to bronze or brass in that it is hard but can be worked when hot, and sharpened to a fine cutting edge which can be resharpened when necessary. The reduction of iron ores is more difficult than for copper, requiring higher temperatures than arise in a simple fire. However, by using bellows to increase the temperature of burning charcoal it is possible to obtain metallic iron from minerals. The smelting and use of metallic iron appears to have been developed in Asia Minor (modern Turkey) by *ca.* 2000 BC. However, this technology did not become widespread until much later, so that the the 'Iron Age' did not really begin until about 1200 BC. Since that time, a huge variety of alloys based on iron has been developed and countless commonplace modern items contain ferrous metal components.

The third major phase of discovery of the transition elements came about during the 18th and 19th centuries. This was stimulated by the increasing understanding of chemical transformations and the improved methods of separation developed by the alchemists. The appearance of Dalton's atomic theory in 1803, followed by the Periodic Table in 1868, gave further impetus to the search for new elements and many new transition elements were discovered during this period.

The fourth phase of discovery came with the more detailed knowledge of atomic and nuclear structure and the discovery of radioactivity, which arose at the end of the 19th and beginning of the 20th centuries. Although cerium had been isolated in 1839, the concept of an f series of elements did not exist at that time. Thus it was not until 1913 that the lanthanide elements were found to constitute a new series. Until then, the similarity in the properties of these elements had led to their mixtures being thought of as single elements. However, in 1918 it was recognized by Bohr that these elements actually constituted the series of 4f elements. Among the actinides, thorium was found in 1829 and uranium in 1789, although uranium metal was not isolated until 1841. The more radioactive 5f elements protactinium, neptunium, plutonium, americium and curium were only discovered in the 20th century. The production of the full series of 6d-block elements has only been achieved in the later part of the 20th century.

Initially, the transition metals presented a chemical puzzle because their halides could form compounds with molecular species such as ammonia, despite the fact that the valencies of all the elements in the metal halide and the ammonia molecule are already satisfied. The problem over the nature of these '*complex compounds*' or complexes of the transition metals led to much debate near the end of the 19th century. The matter was finally resolved through the work of Alfred Werner who was awarded the 1913 Nobel Prize. Werner proposed the concept of a primary valence and a secondary valence for a metal ion (Box 1.2). Werner also showed how the number of compounds formed by a transition metal which have the same formula but different properties, *i.e.* the number of isomers, can reveal structural information. The subsequent theoretical work of Bohr, Schrödinger, Heisenberg and others in the early part of the 20th century provided the basis for a more detailed understanding of the electronic structures of atoms.

Molecules of the same chemical composition, but having different structures, are known as **isomers**. As an example, a metal atom M bonded to four atoms X could have a tetrahedral (1.1) or a square planar (1.2) structure. In order to represent such three-dimensional structures properly in a two-dimensional structural formula, it is necessary to use symbols which provide information about the direction of a bond with respect to the plane of the paper upon which the diagram is drawn. A common convention is to use a *line* to represent a bond in the plane, a *solid wedge* to represent a bond coming up out of the plane, and *dashes* or *ladder wedges* to represent a bond going down into the plane. *Thickly drawn lines* may also be used to represent bonds or lines at the front of a structural diagram.

1.1

1.2

Box 1.2 Primary and Secondary Valence

Binary compounds such as $CrCl_3$, $CoCl_2$ or $PdCl_2$ have valencies of 3, 2 and 2, respectively. Werner proposed that these be called the primary valencies. In a series of cobalt(3+) chloride compounds with ammonia it was found that some of the chorides present could be precipitated as AgCl on adding Ag^+ ions, but that some chloride remained unavailable to this reaction (equations 1.1–1.4).

Yellow: $CoCl_3.6NH_3 + \text{excess } Ag^+ \longrightarrow 3AgCl\downarrow$ (1.1)

Purple: $CoCl_3.5NH_3 + \text{excess } Ag^+ \longrightarrow 2AgCl\downarrow$ (1.2)

Green: $CoCl_3.4NH_3 + \text{excess } Ag^+ \longrightarrow 1AgCl\downarrow$ (1.3)

Violet: $CoCl_3.4NH_3 + \text{excess } Ag^+ \longrightarrow 1AgCl\downarrow$ (1.4)

These observations, together with the results of solution conductivity measurements, can be explained if six groups, either chloride ions or ammonia molecules, remain bonded to the cobalt ion during the reaction and the compounds are formulated as shown in Table 1.4, where the atoms within the square brackets form a single entity which does not dissociate under the reaction conditions. Werner proposed the term secondary valence for the number of groups bound directly to the metal ion; in these examples the secondary valences are six in each case.

Table 1.4 Formulations of cobalt(+3) chloride–ammonia complexes

Colour	*Formula*	*Solution conductivity*
Yellow	$[Co(NH_3)_6]^{3+}3Cl^-$	shows 3:1 electrolyte
Purple	$[CoCl(NH_3)_5]^{2+}2Cl^-$	shows 2:1 electrolyte
Green	$[CoCl_2(NH_3)_4]^+Cl^-$	shows 1:1 electrolyte
Violet	$[CoCl_2(NH_3)_4]^+Cl^-$	shows 1:1 electrolyte

The directionality in the bonding between a d-block metal ion and attached groups such as ammonia or chloride can now be understood in terms of the directional quality of the d orbitals. In 1929, Bethe described the crystal field theory (CFT) model to account for the spectroscopic properties of transition metal ions in crystals. Later, in the 1950s, this theory formed the basis of a widely used bonding model for molecular transition metal compounds. The CFT ionic bonding model has since been superseded by ligand field theory (LFT) and the molecular orbital (MO) theory, which make allowance for covalency in the bonding to the metal ion. However, CFT is still widely used as it provides a simple conceptual model which explains many of the properties of transition metal ions.

The interactions between transition elements and organic molecules through metal–carbon bonds to form organometallic compounds also became a topic of intense interest during the 1950s. In due course, as research in this area reached fruition, it resulted in the award of two Nobel prizes. The first went to K. Ziegler and G. Natta in 1963 for their work on the polymerization of alkenes using organometallic catalysts. The second was awarded in 1973 to G. Wilkinson and E. O. Fischer for their work on organometallic chemistry, in particular on 'sandwich compounds', or metallocenes, in which a transition metal forms the filling between two planar cyclic alkenes in a sandwich-like structure (Box 1.3).

An **organometallic compound** is one which contains one or more metal-to-carbon bonds, although compounds containing bonds to cyanide are not usually included.

Box 1.3 Metallocenes

Metallocenes are sometimes referred to as 'sandwich compounds' because the metal ion is sandwiched between two planar cyclic alkenes as shown in Figure 1.2. Although metallocenes had been synthesized some years earlier, their true structures were not established until the structure of ferrocene was reported in 1951.

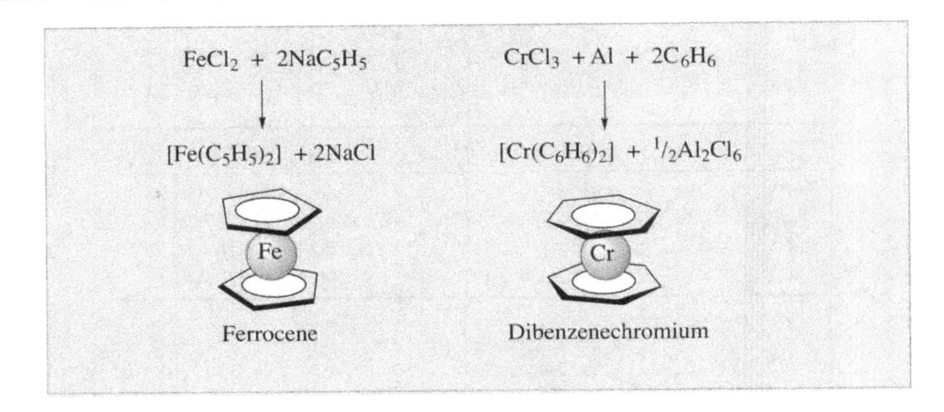

Figure 1.2 Two examples of metallocene compounds

A **metal cluster** has been defined as a compound in which two or more metal atoms are directly linked and in which metal–metal bonding makes a significant contribution to the enthalpy of formation. Examples are shown in Figure 1.3.

Another topic of great interest in the 1950s concerned the mechanism of transfer of electrons between transition metal ions in solution. Work on this topic earned H. Taube a Nobel prize in 1983. Prior to this, the 1981 prize was awarded to K. Fukui and R. Hoffmann for their work on theoretical models of bonding and reactivity, which included studies of transition element compounds. The ability of transition metals to bond to one another directly has provided another active area of research. This has provided examples of metal clusters containing from two up to hundreds of metal atoms linked by metal–metal bonds. Chemists can now investigate the point at which a group of metal atoms becomes sufficiently small so that it ceases to behave like a metallic material and assumes the properties of a molecular entity.

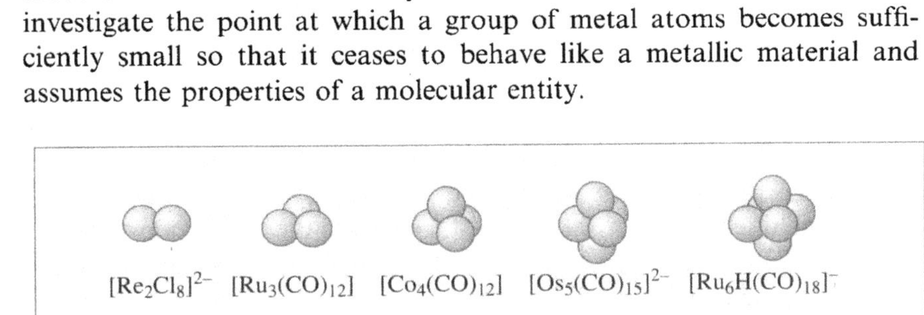

Figure 1.3 Clusters containing two, three, four, five and six metal atoms

Towards the end of the second millennium, studies of the transition elements continued to make major contributions to chemical science and technology. The development of new catalysts and reagents represents one area of activity. Examples are provided by the activation of saturated hydrocarbons by rhodium or lutetium complexes, new syntheses of optically active products in reactions which employ chiral metal compounds, and transition metal compounds which catalyse the stereospecific polymerization of alkenes. The ability of transition metal centres to bind to several organic molecules has been exploited in the construction of new two- and three-dimensional molecular architectures (Figure 1.4). New materials containing transition elements are being developed, one

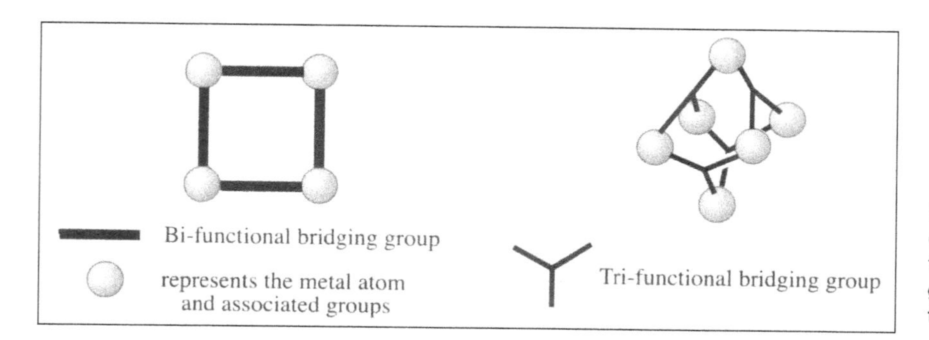

Figure 1.4 Assembled molecules containing four metal centres with four bifunctional bridging groups and six metal atoms with four trifunctional bridging groups

famous example being the high-temperature ceramic superconductors which contain copper and lanthanide ions.

When cooled to very low temperatures, some materials lose all electrical resistance and exhibit **superconductivity**. Some alloys are superconductive when cooled in liquid helium (boiling point 4.2 K), but some ceramics are superconductive at higher temperatures and need only be cooled in liquid nitrogen (boiling point 77.3 K) to show this effect.

Another recent discovery is that of collosal magnetoresistance (CMR), in which the electrical resistance of a material changes dramatically in the presence of a magnetic field. This property is important in the devices which read data on magnetic storage devices, such as computer hard discs.

The late 20th century has also seen a remarkable growth in the understanding of biological systems containing transition metal ions. The crystal structures of many metalloproteins have been determined and even the structure of a nitrogenase enzyme, which has been sought for many years, is now known. This remarkable compound uses a metal cluster (**1.3**) containing iron and molybdenum to convert atmospheric nitrogen to ammonia under ambient conditions. The Haber–Bosch process used industrially requires an iron catalyst, temperatures of about 450 °C and a pressure of about 200 atm (152,000 Torr) to effect this reaction. Clearly chemists have much to learn from the study of metalloproteins.

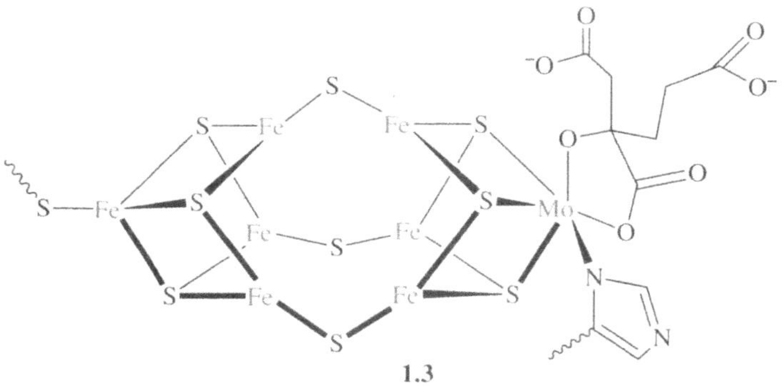

1.3

1.4 Some Applications of the Transition Elements

The particular properties of the transition elements are exploited in a remarkable variety of applications. Some metals are used in very large quantities, particularly iron in structural materials, while others are used in only small quantities for specialized applications such as the synthesis of fine chemicals.

As well as having electrical conductivity, the transition elements can be used in the production of electrical energy through their chemical reactivity. Perhaps the most immediately familiar example is the 'dry cell' battery. Any of a number of chemical reactions may be exploited in this context. As a consequence, manganese, nickel, zinc, silver, cadmium or mercury may be found in dry cells.

Worked Problem 1.2

Q Write chemical equations for the reactions utilized by (i) zinc chloride and (ii) alkaline manganese non-rechargable dry cell batteries.

A An answer to this question may be found by consulting the *Kirk-Othmer Encyclopaedia of Chemical Technology*.

(i) Zinc chloride:

$$4Zn + 8MnO_2 + ZnCl_2 + 9H_2O \rightarrow 8MnOOH + ZnCl_2.4ZnO.5H_2O$$

(ii) Alkaline manganese:

$$2Zn + 3MnO_2 \rightarrow 2ZnO + Mn_3O_4$$

A **nuclear magnetic resonance** or NMR spectrum reveals the energies at which the nuclei of a particular element in a compound attain a resonant condition in a magnetic field and can provide detailed information about chemical structure. This technique can also be used to obtain images of internal organs in the body by **magnetic resonance imaging** or MRI.

The magnetic properties of transition metals are also of great commercial importance. A commonplace example is provided by magnetic recording media such as floppy discs or hard discs and magnetic recording tapes, the coating of which contains metal oxides, typically CrO_2 or γ-Fe_2O_3. Small high-intensity permanent magnets are important in the construction of compact powerful electric motors, such as those used to power windows in cars. Compounds such as $Nd_2Fe_{14}B$ possess these special magnetic properties. A quite different exploitation of the magnetic properties of metal ions is provided by the use of lanthanide ions, especially Eu^{3+} to separate signals obtained in NMR measurements and Gd^{3+} to enhance the contrast of images obtained by MRI.

The colours associated with some transition metal compounds make them useful as pigments. Examples include manganese violet, chrome yellow, cobalt blue, cadmium yellow and Prussian blue, an iron cyanide compound. In some cases it is the absence of colour which is important. Hence, TiO_2 is widely used because of its bright white appearance and is a common component of paints. In other cases it is the ability of an excited metal ion to emit light of a particular frequency which is useful. The lanthanide elements in particular show strong luminescence of the type needed in the cathode ray tubes of colour television sets.

The special chemical reactivity of the transition metals can be exploited in a variety of catalytic processes. One example familiar to most people is the heterogeneous catalyst used in car exhausts. This contains a platinum/rhodium alloy supported on a ceramic matrix and converts the mixture of oxygen, carbon monoxide, hydrocarbons and oxides of nitrogen in the exhaust gases to water, carbon dioxide and nitrogen, all of which are naturally present in the atmosphere. Another catalyst which you may come across in the home is CeO_2. This is a component of the coating on some self-cleaning oven walls. It is present to promote the oxidation in air of the organic deposits formed on the oven wall during cooking. A number of important industrial processes utilize transition metals and a long-established example is provided by the use of palladium chloride with copper chloride in the Wacker process which converts ethene to ethanal (acetaldehyde).

An important modern example of homogeneous catalysis is provided by the Monsanto process in which the rhodium compound **1.4** catalyses a reaction, resulting in the addition of carbon monoxide to methanol to form ethanoic acid (acetic acid). Another well-known process is hydroformylation, in which the reaction of carbon monoxide and hydrogen with an alkene, $RCH{=}CH_2$, forms an aldehyde, RCH_2CH_2CHO. Certain cobalt or rhodium compounds are effective catalysts for this reaction. In addition to catalytic applications, non-catalytic stoichiometric reactions of transition elements now play a major rôle in the production of fine organic chemicals and pharmaceuticals.

Some f-block elements also find uses in organic synthesis. Samarium diiodide is a useful one-electron reducing agent capable of cleaving carbon–halogen or carbon–hydroxyl bonds, and Ce^{4+} compounds find use in oxidation or oxidative coupling reactions. In recent years the study of transition metal-mediated reactions of organic compounds has become a highly important area of commercial chemical research.

A number of the transition metals are essential trace elements for living organisms, so that one medical application of transition metal compounds is in the treatment of deficiency diseases. Particular examples are provided by preparations containing iron to treat anaemia and the use of dietary supplements containing cobalt in the form of vitamin B_{12}, shown as its cyanide derivative in **1.5**. The reverse of this type of treatment involves the removal of excess metal ions from the body using com-

Atoms and molecules can become electronically excited when exposed to light of the appropriate energy. The excited states formed usually lose energy through thermal vibrations and other means, and very quickly return to the original ground state. However, in some cases the excited state can relax back to the ground state by emitting visible light in a process known as **luminesence**. In cathode-ray TV tubes, electrons are used to excite ions which then emit light of a particular frequency. As examples, red light is emitted from excited Eu^{3+} ions and green light from excited Tb^{3+} ions.

Two important types of catalyst may be identified. Firstly, **heterogeneous catalysts** are in a different phase from the reactants. Typically the catalyst might be in the solid phase and the reactants in the gaseous or liquid phase. Secondly, **homogeneous catalysts** operate in the same phase as the reactants, typically all being in solution.

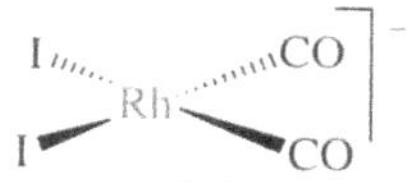

1.4

pounds, sometimes called sequestering agents, which selectively and strongly bind to particular metal ions. Important examples of this are provided by the removal of excess copper to treat Wilson's disease, and the removal of excess iron from patients receiving repeated blood transfusions. Other important therapeutic uses of transition metal compounds include the treatment of cancer using platinum drugs such as carboplatin (**1.6**) and rheumatoid arthritis using gold compounds such as auranofin (**1.7**).

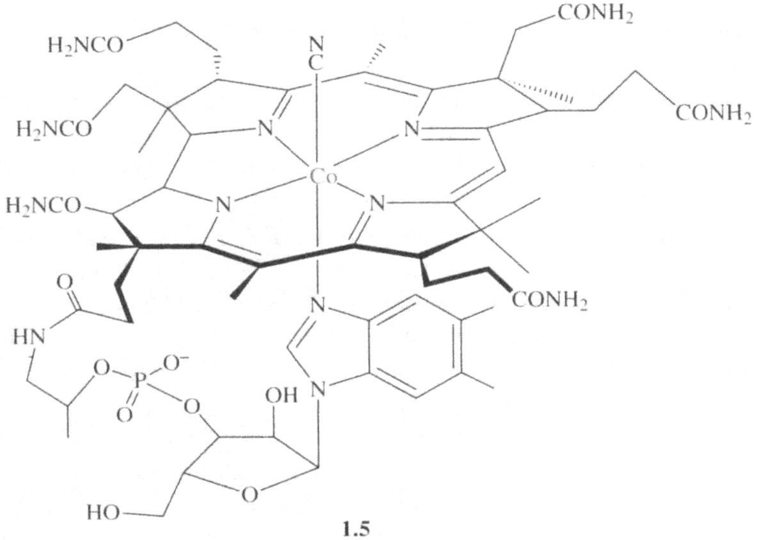

1.5

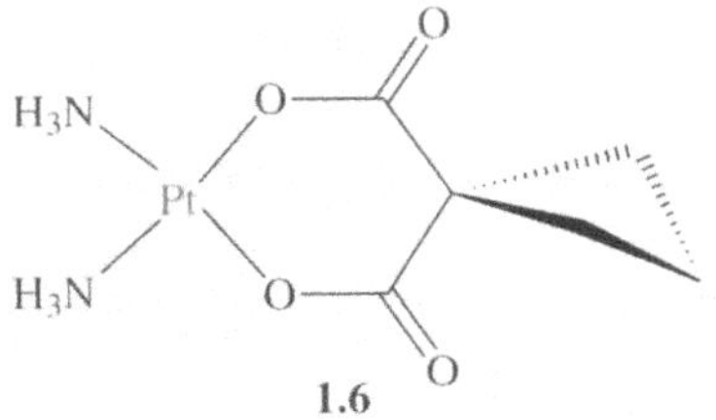

1.6

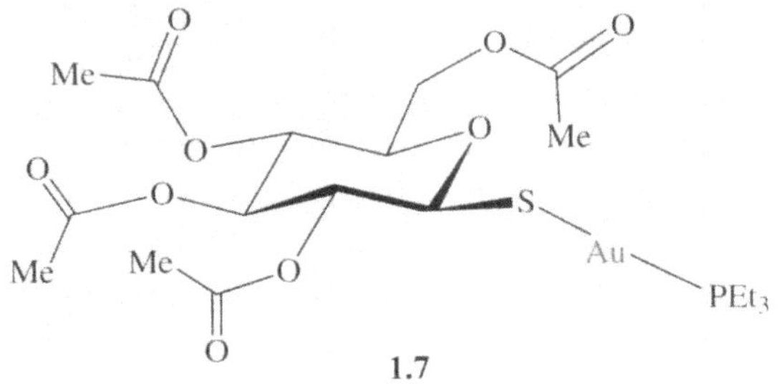

1.7

The transition elements also have applications in diagnostic medicine. The most important element in this context is technetium, specifically the metastable isotope, ^{99m}Tc, which emits a γ-ray of suitable energy for detection by external imaging equipment. This allows the distribution of a ^{99m}Tc complex within the body to be determined and a three-dimensional image created by computer. The development of this important branch of non-invasive diagnostic medicine has only become possible through a detailed understanding of the chemistry of transition elements.

Summary of Key Points

1. *The d- and f-block elements* originate from nuclear reactions in the cosmos and their terrestrial distribution reflects their chemical properties.

2. The *historical development of d- and f-block element chemistry* also reflects their various chemical and physical properties. Metals occurring in elemental form were discovered early in human history, whereas chemically similar elements such as the lanthanides have only relatively recently been identified as separate elements.

3. *The applications of the transitional elements* are many and varied, again reflecting the chemical and physical properties of each particular element. Diverse applications, *e.g.* in catalysts, superconductors and pharmaceuticals, are found for these metals.

Problems

Refer to the further reading section at the end of the book and consult the textbooks suggested, plus any others, to obtain the information needed to complete the following tasks.

1.1. In the answer to Worked Problem 1.1 shown in Scheme 1.1, what is the function of the following reagents:

(i) The $FeCl_2$ in the left branch.
(ii) The $PbCO_3$ in the right branch.
(iii) The $NaHSO_4$ in the right branch.
(iv) The Na_2O_2 in the right branch.
(v) The NaOH treatment of the OsO_4 after its separation from the right branch.

1.2. Consult Chapter 30 in *Advanced Inorganic Chemistry*, 5th edn., by F. A. Cotton and G. Wilkinson, and describe:

(i) Three biochemical processes which require metalloproteins containing iron.

(ii) Three biochemical processes which require metalloproteins containing zinc.
(iii) Three biochemical processes which require metalloproteins containing copper.

1.3. Why are iron or copper ions used in some biological processes rather than zinc?

1.4. On the basis of the following observations made with aqueous solutions, assign primary and secondary valencies to the following metal compounds:

	Formula	Moles of AgCl precipitated with excess added $AgNO_3$
(i)	$PdCl_2.4NH_3$	2
(ii)	$NiCl_2.6H_2O$	2
(iii)	$PtCl_4.2HCl$	0
(iv)	$CoCl_3.4NH_3$	1
(v)	$PtCl_2.2NH_3$	0

2
Atomic Structures and Properties

Aims

By the end of this chapter you should understand the terms:

- Shielding
- Effective nuclear charge
- Ionization energy
- Electronegativity
- Ionic radius
- Oxidation state
- Lanthanide contraction

The chapter will assume some prior knowledge of the principles of atomic structure, including quantum numbers, wavefunctions, principle shells and subshells and the radial distribution function, which indicates the probability of finding an electron as a function of distance from the nucleus.

2.1 Introduction

Under normal conditions, a chemical reaction involves the electrons occupying the outermost shells, or valence shells, of the atoms involved. Hence the chemical properties of an atom arise from its tendency to lose electrons from, or to attract electrons to, its valence shell. This tendency will depend upon the electronic structure of the atom and the nuclear charge experienced by the valence shell electrons. Thus, in order to explain the chemistry of a transition element, it is first necessary to consider its atomic structure and how this influences the binding of its valence shell electrons.

2.2 Transition Elements: Atomic Structures and Properties

2.2.1 Electron Configurations

The square of the radial part of the wavefunction of an orbital provides information about how the electron density within the orbital varies as a function of distance from the nucleus. These **radial distribution functions** show that, in a given principle shell, the maximum electron density is reached nearer to the nucleus as the quantum number l increases. However, the proportion of the total electron density which is near to the nucleus is larger for an electron in an s orbital than in a p orbital.

In the hydrogen atom the s, p, d and f subshells within any principle quantum shell are all of equal energy since only one electron is present. However, in a multi-electron atom the electrons repel one another and so modify the field around the nucleus. Those electrons which spend most of their time nearer to the nucleus shield from the nuclear charge those which, on average, spend most of their time further away. This effect is known as **shielding** and decreases the nuclear charge experienced by an electron in the valence shell of an atom. The nuclear charge actually experienced by an electron is known as the **effective nuclear charge**, Z_{eff}, and, for a valence shell electron, Z_{eff} will be less than the atomic number, Z. The extent to which an electron in an orbital is shielded by the electrons in other subshells depends upon the extent to which it **penetrates** the core electron cloud around the nucleus. The more penetrating an orbital the closer its electrons can, on average, get to the nucleus, and the higher the value of Z_{eff} they experience. Plots of the radial charge distribution functions of orbitals may be used to illustrate these effects. In Figure 2.1a the variation in charge density with distance from the nucleus is shown for the filled core $5s^2$ and $5p^6$ subshells and the partly filled 4f subshell of the $4f^2$ ion Pr^{3+}. These show that the $4f^2$ subshell penetrates the $5s^2$ and $5p^6$ subshells and is not shielded well from the nuclear charge by them. In fact the 4f electrons are 'core like', despite being part of the valence shell, because they are buried within the $5s^2$ and $5p^6$ [Xe] core of the Pr^{3+} ion. In the corresponding early actinide ion U^{4+} the $5f^2$ subshell penetrates the $6s^2$ and $6p^6$ subshells much less and is nearer to the surface of the [Rn] core (Figure 2.1b). This allows more f orbital participation in bonding for the early actinides than for their lanthanide counterparts.

These shielding and penetration effects are apparent in the way the relative energies of the orbitals in atoms vary with increasing atomic number. In any principle shell, the s subshell is more penetrating than the p subshell, so an electron in the s orbital feels a higher Z_{eff}, and has the lower energy. As the atomic number, Z, increases the added core s and p electrons shield the outermost electrons, reducing the rate at which the Z_{eff} they experience increases. In building up the elements from hydrogen to calcium at $Z = 20$, the unoccupied 3d subshell is effectively shielded from the increasing nuclear charge by the electrons of [Ar] core. As a consequence, the energy of the 3d orbitals remains fairly constant . In contrast, the energies of the 4s and 4p orbitals are declining as they penetrate the [Ar] core more. However, the 3d orbitals penetrate

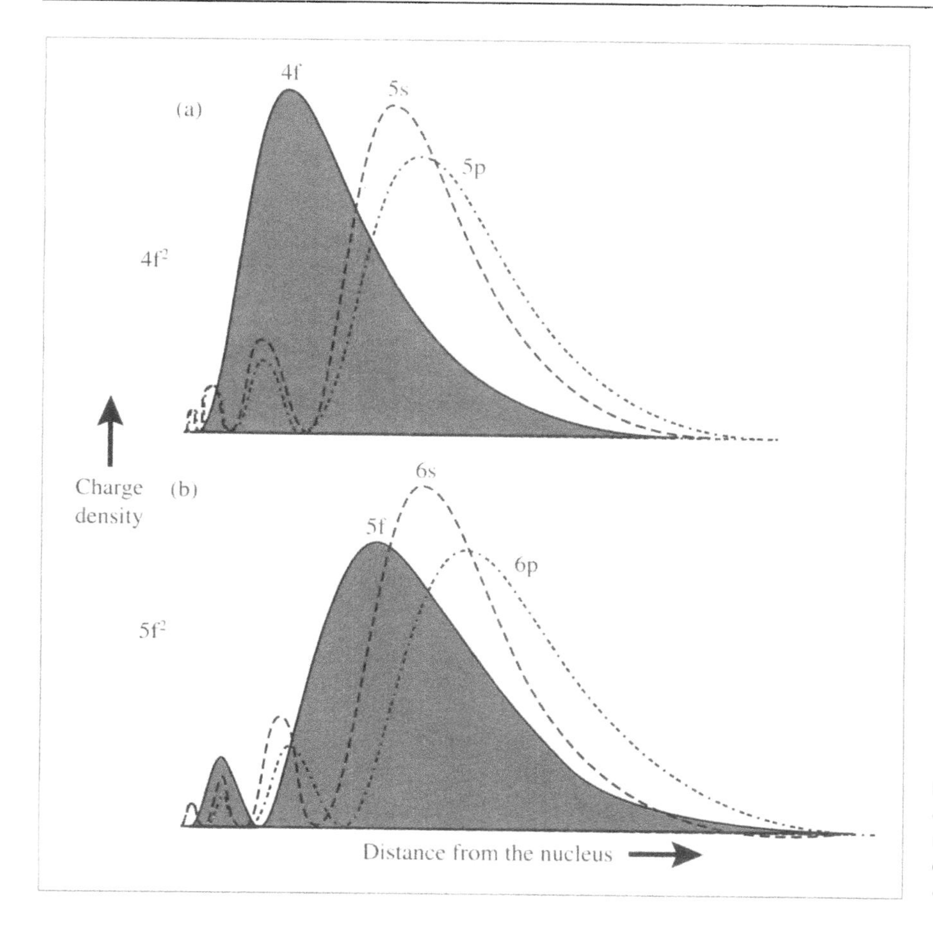

Figure 2.1 Plots of electron density against distance from the nucleus for (a) 4f, 5s and 5p orbitals and (b) for 5f, 6s and 6p orbitals

the 4s and 4p subshells so that after Ca a 3d electron is not fully shielded from the increasing nuclear charge by the $4s^2$ subshell. Thus by the time scandium ($Z = 21$) is reached the 3d subshell has fallen to an energy close to that of the 4s subshell.

Because electrons in a d subshell shield one another from the nuclear charge rather poorly, the energy of the 3d orbitals continues to fall as Z_{eff} increases. Thus, by the time gallium ($Z = 31$) is reached the now filled 3d subshell has fallen well below the 4s and 4p subshells in energy and is no longer part of the valence shell, but has become a core subshell. This pattern is repeated with the 5s, 4d and 5p subshells between $Z = 39$ and 48. However, once barium is reached at $Z = 56$, the 4f, 5d and 6s subshells have converged in energy. As a consequence, after lanthanum, the lanthanide series is formed through the filling of the 4f subshell before the filling of the 5d subshell is completed. This pattern is then repeated again after radium, with the convergence in energy of the 5f, 6d and 7s subshells leading into the actinide series, the 5f subshell becoming increasingly core-like as lawrencium is approached. The electronic configurations of the valence shells of the d-block elements are summarized in Table 2.1 and of the lanthanides and actinides in Table 2.2.

Table 2.1 Valence shell electron configurations, first and second ionization energies E^i, atomic radii and some ionic radii of the d-block metals

Element	Sc	Ti	V	Cr	Mn	Fe	Co	Ni	Cu	Zn
Valence electrons[a]	d^1s^2	d^2s^2	d^3s^2	d^5s^1	d^5s^2	d^6s^2	d^7s^2	d^8s^2	$d^{10}s^1$	$d^{10}s^2$
1st E^i [b]	0.631	0.658	0.650	0.653	0.717	0.759	0.758	0.737	0.746	0.906
2nd E^i [b]	1.235	1.310	1.414	1.496	1.509	1.561	1.646	1.753	1.958	1.733
3rd E^i [b]	2.389	2.652	2.828	2.987	3.248	2.957	3.232	3.393	3.554	3.833
r(atom)[c]	144	132	122	118	117	117	116	115	117	125
$r(M^{+})$[d]	–	–	–	–	–	–	–	–	77	–
$r(M^{2+})$[d]	–	86	79	80	83	78	74.5	69	73	74
$r(M^{3+})$[d]	74.5	67	64	61.5	64.5	64.5	61	60	–	–
$r(M^{4+})$[d]		60.5	58	55	53	58.5	53	–	–	–
Element	Y	Zr	Nb	Mo	Tc	Ru	Rh	Pd	Ag	Cd
Valence electrons[e]	d^1s^2	d^2s^2	d^4s^1	d^5s^1	d^6s^1	d^7s^1	d^8s^1	$d^{10}s^0$	$d^{10}s^1$	$d^{10}s^2$
1st E^i [b]	0.616	0.660	0.664	0.685	0.702	0.711	0.720	0.805	0.731	0.868
2nd E^i [b]	1.181	1.267	1.382	1.558	1.472	1.617	1.744	1.875	2.074	1.631
3rd E^i [b]	1.980	2.218	2.416	2.621	2.850	2.747	2.997	3.177	3.361	3.616
r(atom)[c]	162	145	134	130	127	125	125	128	134	148
$r(M^{+})$[d]	–	–	–	–	–	–	–	–	115	–
$r(M^{2+})$[d]	–	–	–	–	–	–	–	86	–	95
$r(M^{3+})$[d]	90		72	69		68	66.5	76		
$r(M^{4+})$[d]		72	68	65	64.5	62	60	61.5		
Element	La	Hf	Ta	W	Re	Os	Ir	Pt	Au	Hg
Valence electrons[f]	d^1s^2	d^2s^2	d^3s^2	d^4s^2	d^5s^2	d^6s^2	d^7s^2	d^9s^1	$d^{10}s^1$	$d^{10}s^2$
1st E^i [b]	0.538	0.654	0.761	0.770	0.760	0.84	0.88	0.87	0.890	1.007
2nd E^i [b]	1.067	1.44	*1.54*	*1.74*	*1.64*	*1.64*		1.791	1.98	1.810
r(atom)[c]	169	144	134	130	128	126	127	130	134	149
$r(M^{+})$[d]	–	–	–	–	–	–	–	–	137	–
$r(M^{2+})$[d]	–	–	–	–	–	–	–	80	–	102
$r(M^{3+})$[d]	103	–	72	–	–	–	68	–	–	–
$r(M^{4+})$[d]	–	71	68	66	63	63	62.5	62.5	–	–

[a]Electron configurations s^xd^y for $[Ar]4s^x3d^y4p^0$

[b]Ionization energies in MJ mol^{-1}. Values in italic are estimated.

[c]Atomic radii in pm (100 pm = 1 Å).

[d]Effective ionic radii in pm for the metal ion and coordination number 6 or, if *italic*, coordination number 4 and a planar geometry (from Shannon[1]). Crystal radii are *ca.* 14 pm larger for cations and *ca.* 14 pm smaller for anions. In calculations, crystal radii and effective ionic radii should not be mixed.

[e]Electron configurations s^xd^y for $[Kr]5s^x4d^y5p^0$

[f]Electron configurations s^xd^y for $[Xe]6s^x4f^{14}5d^y6p^0$ except La which is $[Xe]6s^24f^05d^16p^0$.

Table 2.2 Electronic structures and ionic radii for the lanthanides and actinides

Lanthanides Element	Valence shell[a] M^0	Ionic radius[b] (pm) M^{3+}	Actinides Element	Valence shell[c] M^0	Ionic radius[b] (pm) M^{3+}	M^{4+}
La	$4f^05d^16s^2$	116	Ac	$5f^06d^17s^2$	*112*	—
Ce	$4f^15d^16s^2$	114.3	Th	$5f^06d^27s^2$	—	105
Pr	$4f^35d^06s^2$	112.6	Pa	$5f^26d^17s^2$	*104*	101
Nd	$4f^45d^06s^2$	110.9	U	$5f^36d^17s^2$	*102.5*	100
Pm	$4f^55d^06s^2$	109.3	Np[d]	$5f^46d^17s^2$	*101*	98
Sm	$4f^65d^06s^2$	107.9	Pu	$5f^66d^07s^2$	*100*	96
Eu	$4f^75d^06s^2$	106.6	Am	$5f^76d^07s^2$	*97.5*	95
Gd	$4f^75d^16s^2$	105.3	Cm	$5f^76d^17s^2$	*97*	95
Tb	$4f^95d^06s^2$	104	Bk	$5f^86d^17s^2$	*96*	93
Dy	$4f^{10}5d^06s^2$	102.7	Cf	$5f^{10}6d^07s^2$	*95*	92
Ho	$4f^{11}5d^06s^2$	101.5	Es	$5f^{11}6d^07s^2$	–	–
Er	$4f^{12}5d^06s^2$	100.4	Fm	$5f^{12}6d^07s^2$	–	–
Tm	$4f^{13}5d^06s^2$	99.4	Md	$5f^{13}6d^07s^2$	–	–
Yb	$4f^{14}5d^06s^2$	98.5	No	$5f^{14}6d^07s^2$	–	–
Lu	$4f^{14}5d^16s^2$	97.7	Lr	$5f^{14}6d^17s^2$	–	–

[a]Valence shell for $[Xe]4f^x5d^y6s^z6p^0$.
[b]Effective ionic radii (pm) for coordination number 8 or, if *italic*, 6 (from Shannon[1]). Crystal radii are *ca.* 14 pm larger for cations and *ca.* 14 pm smaller for anions.
[c]Valence shell for $[Rn]5f^x6d^y7s^z7p^0$.
[d]Also cited as $5f^56d^07s^2$.

2.2.2 Ionization of Transition Element Atoms or Ions

It might be expected from the metallic character of the transition elements that ionization would play a dominant rôle in their chemistry. The electron configurations of the first-row d-block elements reveal the presence of a filled [Ar] core within a valence shell consisting of occupied 4s and 3d subshells. The ionization energy, E^i, or ionization enthalpy, IE, required to remove an electron from these valence subshells will depend upon the Z_{eff} experienced by the electrons involved. Plots of Z_{eff} for 4s and 3d valence electrons against atomic number in neutral atoms are shown in Figure 2.2. The plot for a 3d electron rises more steeply than that for a 4s electron. This difference arises because the 3d electrons lie within the 4s subshell. They shield the 4s electrons from the nuclear charge to some extent and reduce the rate at which Z_{eff} for a 4s electron increases with Z. However, since the 3d orbitals penetrate the 4s subshell, they are more exposed to the increasing nuclear charge than the 4s electrons. The plot of first, second and third IEs for the first row of

The **ionization energy** (E^i) of an atom or ion is defined as the molar change in internal energy for the for the process shown in equation 2.1. This involves the removal of an electron to infinity from the surface of each gaseous atom or ion, each atom or ion being in its lowest energy state. The units of E^i are kJ mol^{-1} or, because the numbers are often large, MJ mol^{-1} (1 MJ mol^{-1} = 1000 kJ mol^{-1}). The standard molar enthalpy change for the same process is the **ionization enthalpy** (IE), where IE = E^i + RT [T is the absolute temperature (degrees Kelvin) and R the gas constant (8.3145 J K^{-1} mol^{-1})]. Since RT is small at room temperature (*ca.* 2.5 kJ mol^{-1}) compared to IE, the values of IE and E^i for transition elements typically differ by less than 1%. The term **ionization potential** is sometimes used to represent the energy of the ionization process and is expressed in units of eV $atom^{-1}$, where 1 eV $atom^{-1}$ = 96.4869 kJ mol^{-1}.

the d-block is shown in Figure 2.3 and, as might be expected from the plots in Figure 2.2, the third IE increases more rapidly with Z than the first and second IEs.

$$M^{z+}_{(g)} \rightarrow M^{(z+1)+}_{(g)} + e^{-}_{(g)} \tag{2.1}$$

Because, in transition metal ions, the energy of the 3d subshell lies below that of the 4s subshell, the remaining valence shell electrons in 3d-block metal ions occupy the 3d rather than the 4s subshell. The second- and third-row d-block elements show similar general trends of

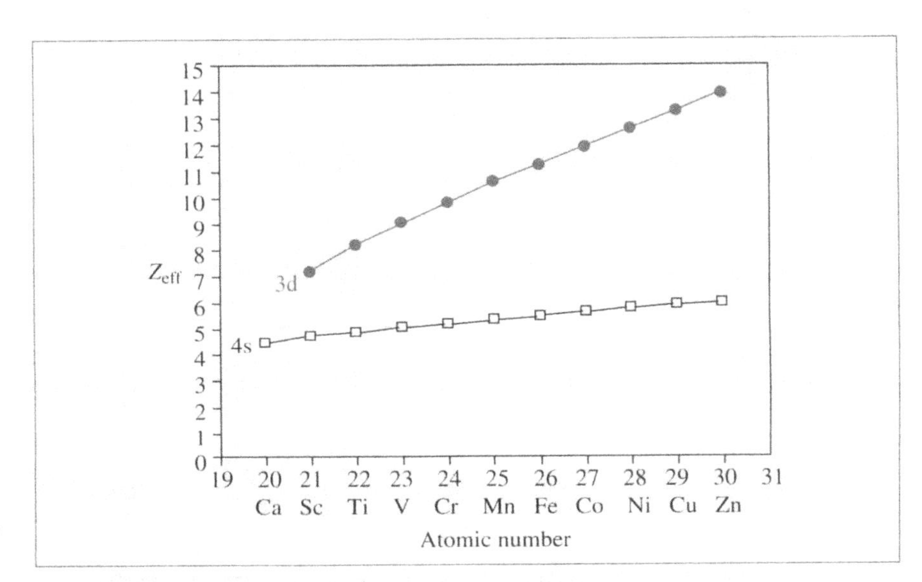

Figure 2.2 The variation in effective nuclear charge (Z_{eff}) with atomic number for electrons in 4s and 3d orbitals

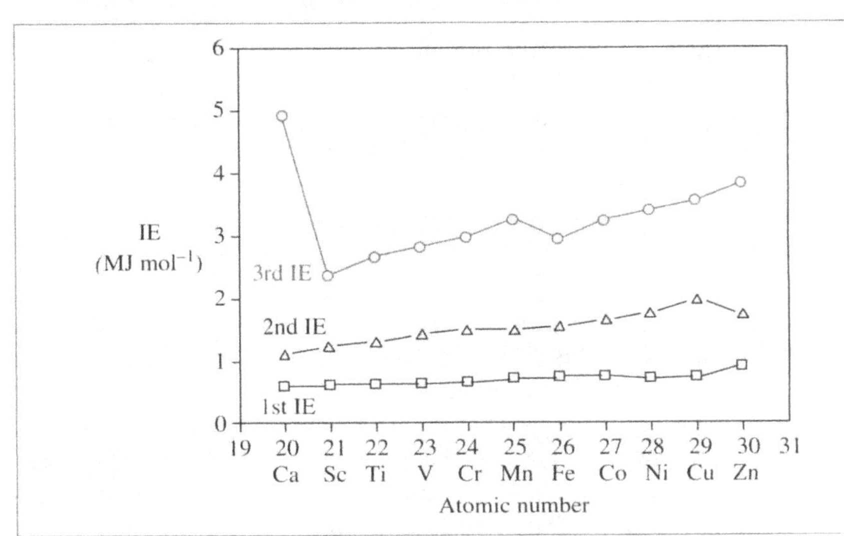

Figure 2.3 The variation in ionization enthalpies (IE) with atomic number for the first row of the d-block

increasing IEs with increasing Z (Table 2.1, Figure 2.4). A particular feature of the plots of third IE against atomic number is the discontinuity between manganese and iron in the first row, and between technetium and ruthenium in the second row. This reflects the additional energy required to break into the half-filled d^5 subshells of Mn^{2+} and Tc^{2+}, which involves the maximum loss of exchange energy (Box 2.1).

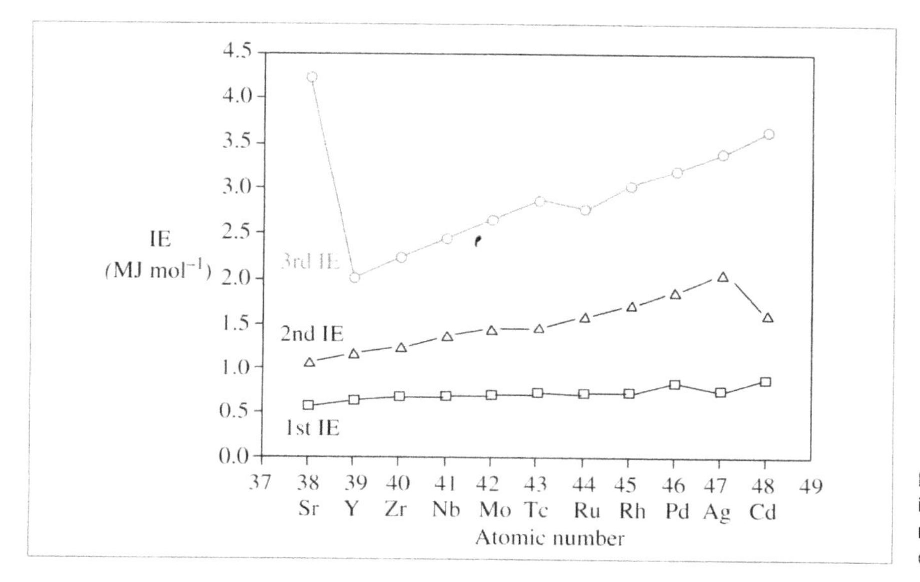

Figure 2.4 The variation in ionization enthalpies (IE) with atomic number for the second row of the d-block

Box 2.1 Exchange Energy

Quantum mechanics shows that if electrons in different orbitals have parallel spins (↑, ↑) they occupy a larger volume of space than if they have antiparallel spins (↑, ↓). Intuitively this would suggest that the electron–electron repulsion will be smaller in the system with parallel spins than in that with antiparallel spins. In fact such systems do show an additional stability due to an exchange energy, E^x. This is, in effect, an expression of Hund's rule.

The magnitude of E^x increases in proportion to $N(N-1)$ (where N is the number of electrons with parallel spins) as shown by equation 2.2, where K^x is the exchange energy associated with one pair. The exact magnitude of K^x changes from element to element and with any ionic charge. The variation in relative exchange energies (E^x/K^x) of d-block ions with number of d electrons, n, is shown in Table 2.3. The values of E^x/K^x increase from 1 to 10 as the first set of parallel spins is introduced, then again from 11 to 20 as a second set of parallel spins is introduced.

Hund's rule states that the lowest energy, or ground, state of an atom or ion will be that with the highest spin multiplicity. That is the state with the most unpaired electrons. As an example, the energy of an atom or ion with three outer electrons will be lower when they have parallel spins (↑, ↑, ↑), in a spin triplet, than when one pair of the electrons has antiparallel spins (↑, ↓, ↑), in a spin doublet.

$$E^x = \sum \frac{N(N-1)}{2} K^x \qquad (2.2)$$

For the ionization process shown in equation 2.1 the relative change in exchange energy for different numbers of d electrons, n, is given in Table 2.4. This pattern of behaviour can be seen in the plots of third IEs in Figures 2.3 and 2.4.

Table 2.3 Relative exchange energy (E^x/K^x) for d^n ions

n	1	2	3	4	5	6	7	8	9	10
$\Sigma N(N-1)/2$	0	1	3	6	10	10	11	13	16	20

Table 2.4 Change in relative exchange energy for $M^{2+}_{(g)} \rightarrow M^{3+}_{(g)} + e^-_{(g)}$

n in d^n configuration of M^{2+}	1	2	3	4	5	6	7	8	9	10
Change in relative exchange energy[a]	0	1	2	3	4	0	1	2	3	4

[a]Relative exchange energy as defined in Table 2.3.

Worked Problem 2.1

Q Why is $ZrCl_4$ the most stable chloride of zirconium, when for palladium it is $PdCl_2$?

A The third and higher IEs of the d-block metals increase with increasing atomic number owing to the larger Z_{eff} making it more energetically unfavourable to attain oxidation states above +2. Furthermore, the d orbitals become more core-like towards the end of the transition series, and so are less effective in stabilizing higher oxidation states through covalent contributions to bonding. These factors combine to make higher oxidation states less accessible to the right of the d-block. In addition, there is a greater loss of exchange energy in creating Pd^{4+} from Pd^{2+} than Zr^{4+} from Zr^{2+}.

As found for the transition metals, the first and second IEs of the lanthanides show little variation with atomic number (Figure 2.5), but the third IE is far less regular. It might be expected from the larger number of electrons involved that the exchange energy consequences of breaking into a half-filled $4f^7$ or filled $4f^{14}$ subshell would be more pronounced. In accord with this, the third IE does increase strongly to Eu^{2+}, where the $4f^7$ configuration is reached. Thereafter it falls back and then increases again on approaching the $4f^{14}$ configuration at Yb^{2+}.

Continuing this pattern, the first and second IEs of the actinides are less sensitive to increasing atomic number than the third IEs (Figure 2.6). However, owing to the similarity in energies between the 6d and 5f subshells, the behaviour of the third IEs is less simple than for the lanthanides, although maxima do appear at Am^{2+} and No^{2+}. The behaviour of the actinide elements is also complicated by the effects of relativity. These result in a contraction of the 7s and 7p orbitals but an expansion and destabilization of the 6d and 5f orbitals. As a consequence, the actinide valence shell 6d and 5f electrons are more easy to ionize than would be predicted by a non-relativistic model.

The theory of relativity predicts that an object in relative motion will show an increase in its mass related in magnitude to its kinetic energy, the mass of the object becoming infinite when the velocity of light is reached. In light atoms the simple Bohr model of an electron orbiting a nucleus leads to radial velocities for the electron which are a small proportion of the velocity of light so that **relativistic effects** are negligible. However, once the heavier elements have been reached, electron radial velocities can be a significant proportion of the velocity of light and the effects of relativity become important.

Figure 2.5 The variation in ionization enthalpies (IE) with atomic number for the lanthanides

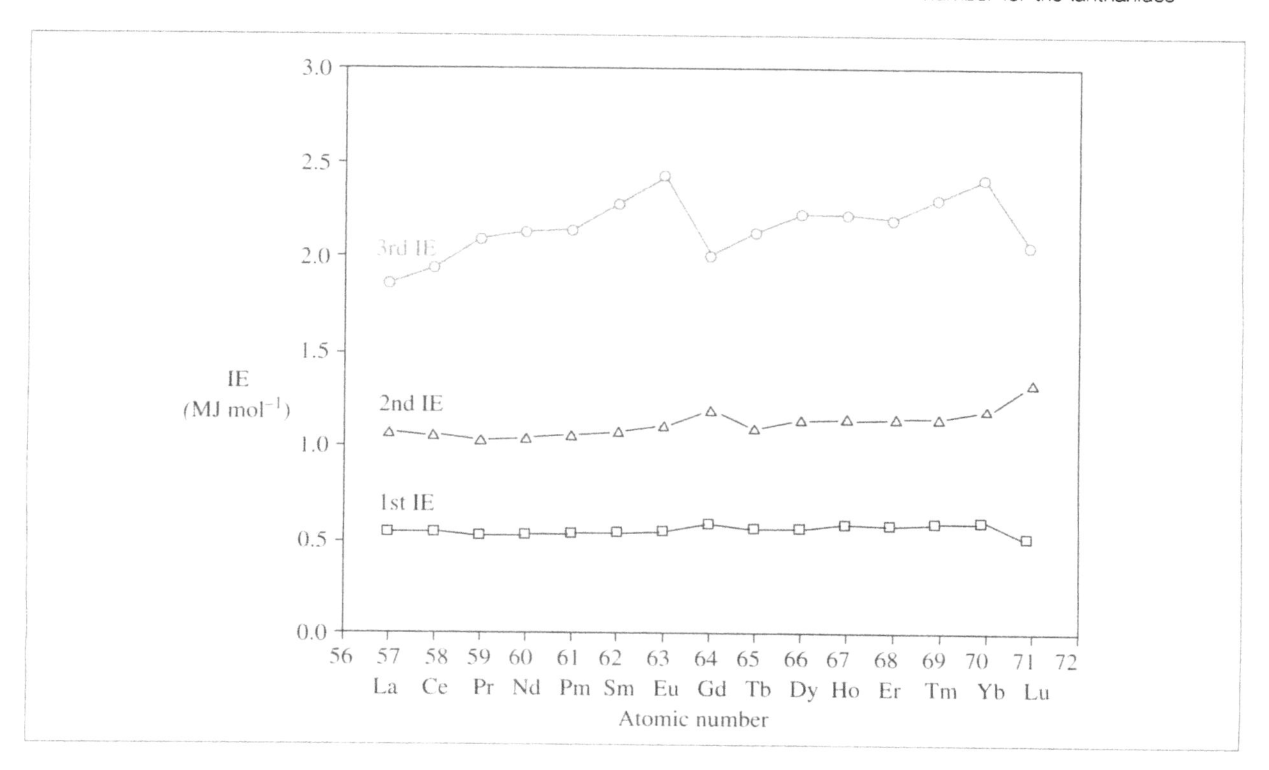

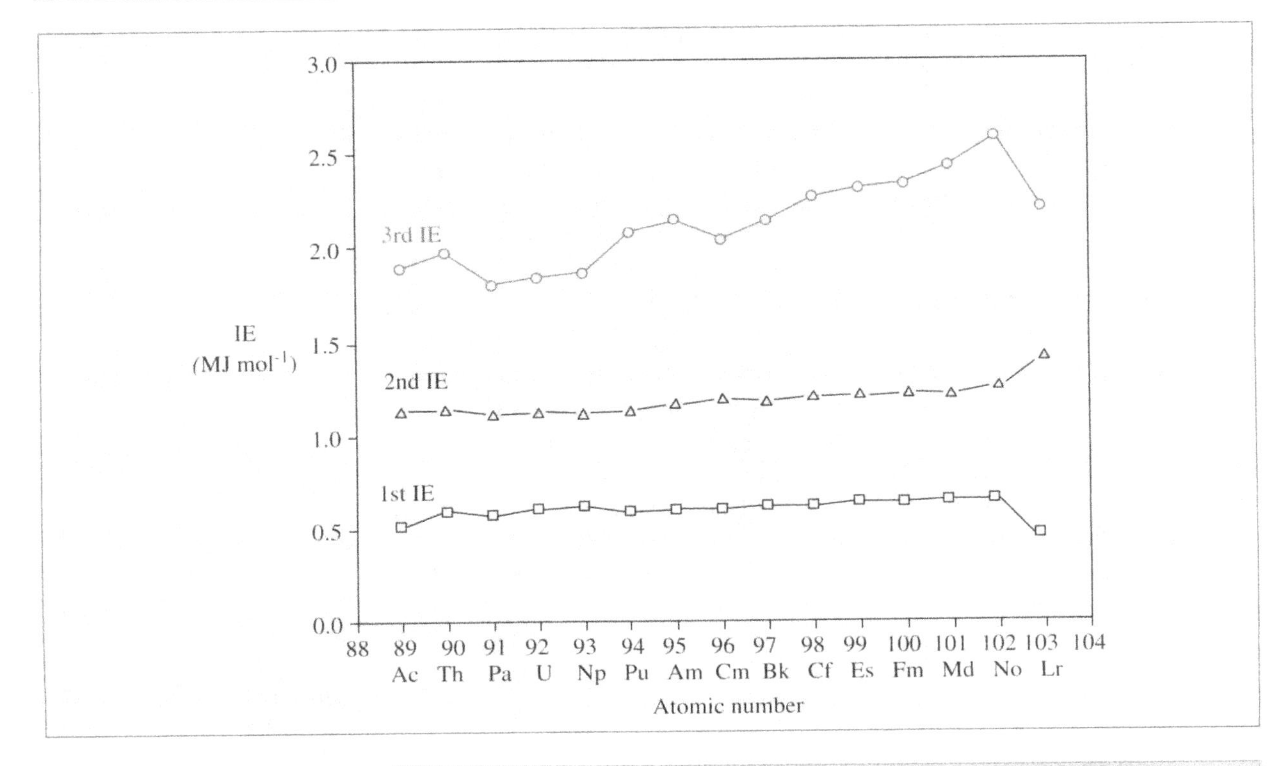

Figure 2.6 The variation in ionization enthalpies (IE) with atomic number for the actinides

Worked Problem 2.2

Q The first two ionization enthalpies of the lanthanide elements increase only slightly with increasing atomic number, but the third increases strongly from La to Eu, then drops back at Gd, only to increase again to Yb and drop back at Lu (see Figure 2.5). Explain these observations.

A After praeseodymium the first two ionizations may be associated with the 6s electrons and are relatively insensitive to Z owing to the shielding effects of the 4f electrons. However, there is a discontinuity in the second IE of gadolinium which has the electron configration $4f^7 5d^1 6s^2$ rather than $4f^8 5d^0 6s^2$, so that it is possible that the first ionization involves the 5d electron and the second breaks into the $6s^2$ subshell of Gd^+. The third IEs show a steady increase associated with an increasing exchange energy loss (Box 2.1), which reaches a maximum on breaking into the half-filled 4f subshell at $4f^7$ Eu^{2+}. After europium there is a discontinuity at gadolinium, for which the third ionization involves removing an electron in excess of the half-filled $4f^7$ subshell, so there is no loss of exchange energy (Table 2.5). Following this discontinuity, the trend seen in the first half of the series is repeated up to $4f^{14}$ Yb^{2+}. Here the third IE involves breaking into the filled $4f^{14}$ subshell with the attendant maximum loss in exchange energy. At lutetium the third IE falls again, as it is associated with removing an electron from outside the closed $4f^{14}$ subshell.

Table 2.5 Relative change in exchange energy for $Ln^{2+}_{(g)} \rightarrow Ln^{3+}_{(g)} + e^-_{(g)}$

n in f^n configuration	1	2	3	4	5	6	7	8	9	10	11	12	13	14
$\Sigma N(N-1)/2$ for f^n	0	1	3	6	10	15	21	21	22	24	27	31	36	42
Relative change in exchange energy for third ionization	0	1	2	3	4	5	6	0	1	2	3	4	5	6

2.2.3 Adding Electrons to Transition Element Atoms

Although ionization plays a dominant rôle in the chemistry of the transition elements, the reverse process of adding an electron to their atoms also contributes to their chemical properties. In fact, adding an electron to the valence shell of most transition elements is an exothermic process. This might be anticipated for elements in which partly filled d or f subshells are present. However, for zinc, cadmium and mercury, which have filled valence shells [$nd^{10}(n+1)s^2$ (n = 3, 4 or 5)], the process of electron addition is endothermic.

The **electron gain energy** (E^a) of an atom or ion may be defined as the standard molar change in internal energy for the process shown in equation 2.3. This involves the attachment of an electron to each gaseous atom or ion in one mole. The usual units of E^a are kJ mol^{-1}. The standard enthalpy change for the same process is the **electron gain enthalpy** (EA), where EA = E^a − RT. The first electron gain enthalpies of neutral atoms may be endothermic or exothermic, depending on the element. However, the second electron gain enthalpies are all endothermic.

The term **electron affinity** has been defined as the difference in internal energy between one mole of gaseous atoms of an element and one mole of its gaseous monoanions. A positive electron affinity indicates that the anions are of lower energy than the atoms and that the formation of the anions from the atoms is exothermic.

Electron Gain Energy, Electron Gain Enthalpy and Electron Affinity

The standard molar energy change associated with the attachment of an electron to an atom or ion according to equation 2.3 is its electron gain energy, E^a. It is perhaps surprising to find that the first E^a values for some transition metals are comparable to those of non-metallic elements.

$$M^{x-}_{(g)} + e^-_{(g)} \rightarrow M^{(x+1)-}_{(g)} \quad (2.3)$$

As examples, the first E^a values of platinum and gold (Table 2.6) are comparable with that of sulfur (−200 kJ mol^{-1}) and larger than that of oxygen (−141 kJ mol^{-1}). The first E^a values of the other platinum metals, except palladium, and of nickel and copper are comparable with the E^a of carbon (–122 kJ mol^{-1}). These observations might lead us to expect that the chemistry of the d-block metals will be rather more subtle than the simple formation of cations as found for the s-block metals.

Table 2.6 Electron gain energies[a] (E^a) and electronegativities[b] (χ) of the transition elements

	Sc	Ti	V	Cr	Mn	Fe	Co	Ni	Cu	Zn
E^a	*30*	*–7.6*	*–51.7*	*–64*	*50*	*–16*	*–64*	*–112*	*–119*	*60*
χ_P	**1.4**	**1.5**	**1.6**	**1.7**	**1.6**	**1.8**	**1.9**	**1.9**	**1.9**	**1.7**
χ_{AR}	1.2	1.3	1.5	1.6	1.6	1.6	1.7	1.8	1.8	1.7
	Y	Zr	Nb	Mo	Tc	Ru	Rh	Pd	Ag	Cd
E^a	*0*	*–41*	*–86*	*–72*	*–53*	*–101*	*–110*	*–54*	*–126*	*70*
χ_P	**1.2**	**1.3**	**1.6**	**2.2**	**1.9**	**2.2**	**2.3**	**2.2**	**1.9**	**1.7**
χ_{AR}	1.1	1.2	1.2	1.3	1.4	1.4	1.5	1.4	1.4	1.5
	La	Hf	Ta	W	Re	Os	Ir	Pt	Au	Hg
E^a	*–50*	*–10*	*–31*	*–79*	*–14*	*–110*	*–151*	*–205*	*–223*	*50*
χ_P	**1.1**	**1.3**	**1.5**	**2.4**	**1.9**	**2.2**	**2.2**	**2.3**	**2.5**	**2.0**
χ_{AR}	1.1	1.2	1.3	1.4	1.5	1.5	1.6	1.4	1.4	1.4
	Ac									
E^a	–									
χ_P	**1.1**									
χ_{AR}	1.0									

	Ce–Nd		Sm–Er		Tm–Lu	
E^a	*–50*		*–50*		*–50*	
χ_P	**1.1**		**1.1**		**1.1**	
χ_{AR}	1.1		1.2		1.3	
	Th	Pa	U	Np	Pu	Am–No
E^a	–	–	–	–	–	–
χ_P	**1.3**	**1.5**	**1.4**	**1.4**	**1.3**	**1.3**
χ_{AR}	1.1	1.1	1.2	1.2	1.2	1.2

[a]Electron gain energies (E^a in kJ mol^{-1}) are shown in *italic* text and are taken from Hotop and Lineburger[2] or Bratsch and Lagowski[3].

[b]Electronegativities, χ_P, shown in **bold** text and rounded to one decimal place, are taken from Allred[4] or from Pauling.[5] The values are based on thermochemical data analysed using Pauling's approach for the elements in their 'normal' oxidation state, *viz.* M^{3+} for Sc, Y and La, M^+ for Cu, Ag and Au, M^{2+} for the other d-block metals, Ln^{3+} for the lanthanides and An^{3+} for the actinides.

[c]Electronegativities, χ_{AR}, shown in normal typeface are calculated using the Allred–Rochow method and are taken from Allred and Rochow[6] or from Little and Jones.[7]

Electronegativity

The chemical properties of an element reflect the net effect of its IE and E^a values, as indicated by its electronegativity, χ (Box 2.2, Table 2.6). Unlike IE and E^a values, which refer to isolated atoms in the gas phase, χ refers to atoms in molecules and so aims to evaluate the overall ability of an atom to attract charge to itself in a real chemical environment. Values of χ can provide a useful insight into the chemical properties of the transition elements. The lighter early d-block metals, the lanthanides and the actinides show the lower Pauling electronegativity, χ_P, values. These fall in the range 1.0–1.5, which may be compared with χ_P values of 0.8–1.0 for the alkali metals and 1.3 for magnesium. The platinum metals, together with molybdenum and tungsten, show larger χ_P values, in the range 2.2–2.4, which may be compared to values of 2.0 for phosphorus, 2.2 for hydrogen and 2.6 for carbon. This suggests that the polarity of bonds between these metals and elements such as carbon, hydrogen and phosphorus will be lower than for the early d-block metals, the lanthanides or the actinides. In other words, the contribution of covalency to bonding in platinum metals, molybdenum or tungsten compounds will be greater.

Box 2.2 Electronegativity

Pauling originally defined electronegativity, χ, as "The power of an atom in a molecule to attract electrons to itself". In covalent compounds, heteronuclear bonds between dissimilar atoms are usually stronger than homonuclear bonds between atoms of the same element. These differences in bond energy may be assigned to the increased contribution from ionic bonding as the difference in electronegativity between the two atoms increases. On this basis, Pauling used thermochemical data to devise a scale of electronegativities for the elements, known as Pauling electronegativities, χ_P. Subsequently, Allred used more recent thermochemical data to recalculate some of Pauling's values and obtain a revised set of χ_P values.

Several other approaches have been proposed for determining electronegativities but only two of these will be mentioned here. Mulliken proposed an electronegativity scale based on taking χ to be the average of the E^i and E^a values of the element. Allred and Rochow proposed that electronegativity be defined as the force exerted by the nucleus on the valence electrons, so that the Allred–Rochow electronegativity, χ_{AR}, can be calculated from the

Z_{eff} and the covalent radius, r, in pm (Section 2.2.4), of an atom using equation 2.4:

$$\chi_{AR} = (3590Z_{eff}/r^2) + 0.744 \quad (2.4)$$

Although these two approaches give similar values for many of the main group elements, rather different values are obtained for some of the d-block metals. Values for both χ_P and χ_{AR} are given in Table 2.6. The χ_P values may be considered to be experimental results in that they are obtained from thermochemical measurements, while the χ_{AR} values are theoretical in that they are obtained by calculation. However, the two scales cannot be directly compared as they are based on different definitions of χ.

Oxidation States

The oxidation state (Box 2.3), or oxidation number, of an element is a concept which is widely used in discussions of transition element chemistry. Strictly speaking, the term oxidation state only has any clear chemical meaning in fully ionic compounds containing elements of very different electronegativity. In KCl, for example, we have no difficulty in assigning oxidation states of +1 to K^+ and –1 to Cl^-. However, in a compound such as MoS_2, where the two elements have very similar electronegativity values (χ = 2.24 for Mo^{4+} and 2.58 for S), the situation is not so clear-cut. The bonding in MoS_2 may involve significant covalency and it is not safe to assume that the molybdenum is actually present as Mo^{4+} even though it has been assigned an oxidation state of +4. Despite such difficulties, oxidation state remains a useful formalism in chemical discussions and provides an aid in electron 'book-keeping'.

Box 2.3 Oxidation State

In assigning an oxidation state to an atom of an element in a compound, it is necessary to assume an extreme ionic bonding model, even though this may not be chemically reasonable. In the case of the transition elements, it is usual to consider the transition element as the least electronegative element in the compound and so assign the other atoms or groups bonded to the metal a negative or, for uncharged groups such as H_2O, NH_3 or CO, zero oxidation state. The total of these numbers when subtracted from the charge on the compound gives the oxidation state of the transition element. Sometimes it is necessary to use a little chemical 'horse sense' to choose a suitable value. A particular problem arises with hydrogen since this may be considered to be present as hydride, H^-, or as a proton, H^+. In such cases you might ask: does the compound reacts with bases to remove H^+ or with protic acids to evolve H_2? Another ambiguous case is provided by oxygen. In MoO_2, for example, is this an oxide containing $2O^{2-}$ and Mo^{4+} or a peroxide containing O_2^{2-} and Mo^{2+}? In salts such as $Na_2[PtCl_4]$ or $[Co(NH_3)_6]Cl_3$ it is necessary to separate the metal complex from the counterions; in these examples this gives $[PtCl_4]^{2-}$ and $[Co(NH_3)_6]^{3+}$, respectively.

Some examples:

WS_2 — Assigning sulfur its normal oxidation state of −2 gives, for a zero charge on the compound: $0 - \{2 \times (-2)\} = +4$ for tungsten, *i.e.* $W^{4+}2S^{2-}$ and so W(+4).

$K_2[PtCl_6]$ — Here the platinum complex is $[PtCl_6]^{2-}$, so assigning a normal oxidation state of −1 to chlorine gives, for a −2 charge on the complex: $-2 - \{6 \times (-1)\} = +4$ for platinum, *i.e.* $Pt^{4+}6Cl^-$ and so Pt(+4).

$Na_2[Fe(CO)_4]$ — Here the iron complex is $[Fe(CO)_4]^{2-}$ and CO is a neutral group giving, for a −2 charge on the complex: $-2 - \{4 \times 0\} = -2$ for iron, *i.e.* a negative oxidation state of −2.

$[Mn(CH_3)(CO)_5]$ — Here the more electronegative carbon means that the methyl group is present as CH_3^- and CO is neutral, as is the complex, so we have: $0 - (-1) - (5 \times 0) = +1$ for manganese, *i.e.* Mn(+1).

Worked Problem 2.3

Q Write down the oxidation state and valence shell electron configuration of the metal in each of the following ions: Ti^{3+}, ZrO^{2+}, RuO_4^{2-}, Ni^{3+}.

A In transition metal ions, all remaining valence shell electrons are assigned to the outer shell d orbital.

Titanium is $[Ar]4s^23d^2$ so Ti^{3+} (oxidation state +3) is $3d^1$
Zirconium is $[Kr]5s^24d^2$ so ZrO^{2+} (oxidation state +4) is $4d^0$
Ruthenium is $[Kr]5s^24d^6$ so RuO_4^{2-} (oxidation state +6) is $4d^2$
Nickel is $[Ar]4s^23d^8$ so Ni^{3+} (oxidation state +3) is $3d^7$

2.2.4 Radii of Atoms and Ions

van der Waals Radii and Covalent Radii

If atoms were hard spheres with well-defined surfaces like billiard balls it would be relatively easy to assign radii to them. However, the electron cloud of an atom is not hard and does not have a well-defined boundary, so the method by which the radius is measured will affect the numerical value obtained. Measuring the distance between atoms in a solid array which appear to be touching, but are not bonded together, would be one approach. This is possible with the noble gases, where the atoms lie with an equilibrium separation at which the attractive van der Waals force between them is balanced by the repulsive force between their closed electron shells. Half the distance between atoms of a noble gas in the solid phase could be taken as its atomic radius. In the case of other elements, the situation is less simple. Usually it is necessary to analyse a large number of atom–atom distances from different structures to make a reasonable estimate of an atomic radius. The average figure obtained from the distances between non-bonded atoms could be used to estimate a van der Waals radius. It is also possible to measure radii from the structures of the elemental forms of the element in question, although these do not necessarily represent a simple non-bonded radius as some bonding interactions may be present. However, this does provide one convenient means of measuring atomic radii for metals. In the case of covalent compounds, a large number of distances between bonded atoms may be measured and a self-consistent set of covalent radii devised.

The **van der Waals radius** of an atom represents the distance from its nucleus at which the attraction due to the van der Waals forces between it and the surface of a neigbouring non-bonded atom is balanced by the repulsive forces between the electron shells surrounding the two atoms. The sum of the van der Waals radii of two non-bonded atoms in contact under these conditions is the internuclear distance between the two atoms.

The **covalent radius** of an atom in a molecule is the distance from its nucleus at which the potential energy of the atom–atom interaction with its neighbour reaches a minimum. In a homonuclear bond the covalent radius of the element is half the internuclear distance. In the case of heteronuclear bonds, the covalent radii will need to be apportioned between the two elements on the basis of analysing a number of bond distances to obtain a self-consistent set of values. The sum of the covalent radii of two bonded atoms in a molecule is their internuclear separation.

Worked Problem 2.4

Q The atomic radii of the lanthanide elements, other than Eu and Yb, decrease from 186 pm at La to 174 pm at Lu, but the respective radii of Eu and Yb are much larger at 209 and 193 pm. Why is this?

A An answer may be found by consulting Greenwood & Earnshaw, *Chemistry of the Elements, 2nd edn.*, page 1234. Generally the radii of the lanthanide metals represent the separation of $4f^n$ Ln^{3+} ions in a 'sea of electrons', or a conduction band made up from the 5d and 6s orbitals. This contains three electrons per lanthanide atom. Lanthanide radii steadily decrease with increasing atomic number, owing to the poor screening of f electrons one by another, and the consequent increase in Z_{eff}. In the cases of Eu and Yb, the third ionization energy is at a maximum (Figure 2.5), so the energy required to promote an electron into the conduction band is increased. This leads to a situation where thermal energy is insufficient to promote the third electron to the conduction band. Thus the radii represent the separation of the much larger Ln^{2+} ions in a 5d/6s conduction band containing only two electrons per lanthanide atom.

Ionic Radii

In the case of metal compounds, ionic radii may be determined by examining the structures of salts containing small, non-polarizable anions. A typical value for the radius of the anion is assumed and the remainder of the inter-ionic distance may then be assigned to the radius of the metal cation. By analysing the crystal structures of many fluoride and oxide, or chloride and sulfide, compounds, Shannon and Prewitt[8] have made a comprehensive study of crystal, or ionic, radii (Tables 2.1 and 2.2). It is important to note that the radii obtained for an ion depend upon on both its oxidation state and its coordination number (CN), that is, the number of atoms or ions in contact with the ion. As a consequence, it would be misleading to compare the radii of different metals unless they were obtained for the same oxidation state and CN. As might be expected, the ionic radius for a particular metal and CN decreases with increasing oxidation state. Similarly, for a given metal and oxidation state the ionic radius increases with increasing CN (Table 2.7).

The **ionic radius** of a monatomic ion in a compound is the distance from its nucleus at which the attraction due to electrostatic forces between it and an adjacent counter ion is balanced by the repulsive forces between the electron shells surrounding the two ions. The sum of the ionic radii of an adjacent cation and anion is the internuclear distance between them.

A common trend in the ionic radii of the transition elements is that they tend to decrease with increasing atomic number in a period. This

Table 2.7 Examples of the effect of coordination number and oxidation state (Roman numerals) on ionic radii (pm)

Metal	*Vanadium*				*Manganese*						*Sm*	*U*
CN	V(+2)	V(+3)	V(+4)	V(+5)	Mn(+2)	Mn(+3)	Mn(+4)	Mn(+5)	Mn(+6)	Mn(+7)	Sm(+3)	U(+4)
4	–	–	–	35.5	66	–	39	33	25.5	25		
5	–	–	53	56	75	58	–	–	–			
6	79	64	58	54	83	64.5	53	–	–	46	95.8	89
7	–	–	–	–	90	–	–	–	–		102	95
8	–	–	–	–	96	–	–	–	–		107.9	100
9	–	–	–	–	–	–	–	–	–		113.2	105
10	–	–	–	–	–	–	–	–	–		124	117

reflects the variation in Z_{eff} with Z. It is particularly pronounced for the lanthanide series, where the shielding of f electrons one by another is relatively poor. As a result, there is a steady decrease in ionic radius (CN 8) from 116 pm for La^{3+} to 97.7 pm for Lu^{3+}. This lanthanide contraction has an important effect on the radii of the third-row d-block elements, as it compensates for the size effect of filling the 5p and 6s subshells between cadmium and hafnium, so that the ionic radii of the third-row d-block metals are very similar to those of the second-row d-block metals (Table 2.8). This contributes to there being a much greater similarity in chemistry between the second- and third-row d-block elements than between the first and second row.

Table 2.8 Effective ionic radii for metal ions (CN 6)

M^{3+}	*r* (pm)	M^{4+}	*r* (pm)	M^{3+}	*r* (pm)	M^{4+}	*r* (pm)
Sc^{3+}	74.5	Ti^{4+}	60.5	V^{3+}	64	Ni^{4+}	48
Y^{3+}	90	Zr^{4+}	72	Nb^{3+}	72	Pd^{4+}	61.5
La^{3+}	103	Hf^{4+}	71	Ta^{3+}	72	Pt^{4+}	62.5

Summary of Key Points

1. *The atomic and ionic properties of the transition elements* underlie their chemical behaviour. The effective nuclear charge experienced by valence shell electrons depends upon shielding and penetration effects.

2. *The ionization enthalpies, electron gain enthalpies and electronegativities* of the transitional elements vary systematically with

their electronic structures and underlie the observed trends in the chemistry of elements within a transition series. Among the d- and f-block metals, exchange energies can make an important contribution to ionization energies and electron affinities

3. *The oxidation state of a metal* is a useful formalism for electron counting but can become ambiguous as covalency in bonding increases.

4. *The van der Waals radius, covalent radius and ionic radius* are important parameters for metallic elements and features such as the lanthanide contraction have important chemical consequences.

Problems

2.1. Explain the appearance of the plots of first, second and third ionization enthalpies for the first-row d-block elements as shown in Figure 2.3.

2.2. Write down the oxidation state and valence shell electron configuration of the metal in each of the following ions:

Cu^+, TaO_4^{3-}, Pr^{3+}, OsO_4, Rh^+, Gd^{3+}, Yb^{2+}, $PaO(OH)^{2+}$, UO_2^{2+}, Pu^{4+}, No^{2+}.

2.3. The effective ionic radii of several d-block metal ions are summarized in Table 2.8. Explain the underlying cause of the pattern of variation among these radii.

2.4. The first electron gain energies of the first-row d-block metals in Table 2.6 become increasingly negative from Sc to Cr, become positive at Mn, become increasingly negative from Fe to Cu and finally become positive at Zn. Explain the underlying cause of this pattern of behaviour.

2.5. (i) Aqueous solutions of Cr^{2+} are rapidly oxidized in air, as are solutions of Fe^{2+}, though more slowly. Explain why aqueous solutions of the intervening ion, Mn^{2+}, should be stable towards aerobic oxidation.

(ii) Although aqueous solutions of Fe^{2+} rapidly undergo irreversible oxidation in air, the metalloprotein haemoglobin reversibly binds dioxygen without becoming oxidized. Briefly explain this difference in behaviour.

2.6. Explain the following observations:

(i) In aqueous solutions the most stable oxidation state of neodymium is +3 but for its actinide counterpart, uranium, oxidation state +6 is most stable.

(ii) In aqueous solutions the most stable oxidation state of ytterbium is +3 but for its actinide counterpart, nobelium, oxidation state +2 is most stable.

References

1. R. D. Shannon, *Acta Crystallogr.*, 1976, **A32**, 751.
2. H. Hotop and W. C. Lineburger, *J. Phys. Chem. Ref. Data,* 1985, **14**, 731.
3. S. G. Bratsch and J. J. Lagowski, *Polyhedron*, 1986, **5**, 1763.
4. A. L. Allred, *J. Inorg. Nucl. Chem.*, 1961, **17**, 215.
5. L. Pauling, *The Nature of the Chemical Bond,* 3rd edn., Cornell University, Ithaca, New York, 1960, p. 93.
6. A. L. Allred and E. G. Rochow, *J. Inorg. Nucl. Chem.,* 1958, **5**, 264.
7. E. J. Little and M. M. Jones, *J. Chem. Educ.*, 1960, **37**, 231.
8. R. D. Shannon and C. T. Prewitt, *Acta Crystallogr.*, 1969, **B25**, 925.

3
Binary Compounds

Aims

By the end of this chapter you should understand the factors determining the stoichiometries of:

- The binary oxides, fluorides and chlorides formed by the transition elements

The chapter will assume an understanding of Hess' law and the thermodynamic terms enthalpy of formation and free energy, together with some prior knowledge of the structures of ionic solids in terms of the close packing of spheres.

3.1 Introduction

The atomic and ionic properties of an element, particularly IE, ionic radius and electronegativity, underly its chemical behaviour and determine the types of compound it can form. The simplest type of compound an element can form is a binary compound, one in which it is combined with only one other element. The transition elements form binary compounds with a wide variety of non-metals, and the stoichiometries of these compounds will depend upon the thermodynamics of the compound-forming process. Binary oxides, fluorides and chlorides of the transition elements reveal the oxidation states available to them and, to some extent, reflect trends in IE values. However, the IEs of the transition elements are by no means the only contributors to the thermodynamics of compound formation. Other factors such as lattice enthalpy and the extent of covalency in bonding are important. In this chapter some examples of binary transition element compounds will be used to reveal the factors which determine the stoichiometry of compounds.

Lattice enthalpy, $\Delta H^\ominus_L$, may be defined as the standard enthalpy change when one mole of a solid is formed from its gaseous constituent ions. Since this will be an exothermic process, $\Delta H^\ominus_L$ will have a negative value. **Lattice energy** differs from lattice enthalpy by a multiple of RT and is equal to $\Delta H^\ominus_L$ when $T = 0$.

3.2 Binary Oxides, Fluorides and Chlorides

3.2.1 Maximum Oxidation States

The highest oxidation state available to an element is usually found among its compounds with the two most electronegative elements, fluorine or oxygen, so that an examination of the binary fluorides and oxides of the transition elements should reveal their maximum chemically attainable oxidation states. The stoichiometric oxides of the d-block metals are summarized in Table 3.1, and the fluorides in Table 3.2. Binary compounds with the less electronegative element chlorine might be expected to show a slightly different range of oxidation states and, for comparison, chlorides are summarized in Table 3.3.

Table 3.1 Simple binary transition metal oxides[a]

Sc	Ti	V	Cr	Mn	Fe	Co	Ni	Cu	Zn
								Cu_2O	
	TiO	VO		MnO	FeO	CoO	NiO	CuO	ZnO
Sc_2O_3	Ti_2O_3	V_2O_3	Cr_2O_3	Mn_2O_3	Fe_2O_3	Co_3O_4			
	TiO_2	VO_2	CrO_2	MnO_2			NiO_2		
		V_2O_5							
			CrO_3						
				Mn_2O_7					
Y	**Zr**	**Nb**	**Mo**	**Tc**	**Ru**	**Rh**	**Pd**	**Ag**	**Cd**
								Ag_2O	
		NbO					PdO	AgO	CdO
Y_2O_3						Rh_2O_3	Pd_2O_3		
	ZrO_2	NbO_2	MoO_2	TcO_2	RuO_2	RhO_2	PdO_2		
		Nb_2O_5	Mo_2O_5						
			MoO_3	TcO_3					
				Tc_2O_7					
					RuO_4				
La	**Hf**	**Ta**	**W**	**Re**	**Os**	**Ir**	**Pt**	**Au**	**Hg**
									HgO
La_2O_3				Re_2O_3		Ir_2O_3	Pt_2O_3	Au_2O_3	
	HfO_2	TaO_2	WO_2	ReO_2	OsO_2	IrO_2	PtO_2		
		Ta_2O_5		Re_2O_5					
			WO_3	ReO_3	OsO_3	IrO_3	PtO_3		
				Re_2O_7					
					OsO_4				

[a]Except for Co_3O_4, oxides with non-integer oxidation states have been excluded

Table 3.2 Simple binary transition metal fluorides[a]

Sc	Ti	V	Cr	Mn	Fe	Co	Ni	Cu	Zn
		VF_2	CrF_2	MnF_2	FeF_2	CoF_2	NiF_2	CuF_2	ZnF_2
ScF_3	TiF_3	VF_3	CrF_3	MnF_3	FeF_3	CoF_3			
	TiF_4	VF_4	CrF_4	MnF_4					
		VF_5	CrF_5						
			CrF_6						
Y	**Zr**	**Nb**	**Mo**	**Tc**	**Ru**	**Rh**	**Pd**	**Ag**	**Cd**
								AgF	
							PdF_2	AgF_2	CdF_2
YF_3	ZrF_3	NbF_3	MoF_3		RuF_3	RhF_3			
	ZrF_4	NbF_4	MoF_4		RuF_4	RhF_4	PdF_4		
		NbF_5	MoF_5		RuF_5	RhF_5			
			MoF_6	TcF_6	RuF_6	RhF_6			
La	**Hf**	**Ta**	**W**	**Re**	**Os**	**Ir**	**Pt**	**Au**	**Hg**
									Hg_2F_2
									HgF_2
LaF_3						IrF_3		AuF_3	
	HfF_4			ReF_4	OsF_4		PtF_4		
		TaF_5		ReF_5	OsF_5	IrF_5	PtF_5	AuF_5	
			WF_6	ReF_6	OsF_6	IrF_6	PtF_6		
				ReF_7					

[a]Fluorides with non-integer oxidation states have been excluded

These tables reveal that, towards the left of the transition series, elements can attain their maximum possible oxidation state or group oxidation state, while at the right of the series oxidation state +2 becomes dominant. In part this reflects the increase in third and higher IEs with increasing atomic number across the transition series and the increasingly 'core-like' nature of the d orbitals. In the first row of the d-block the range over which the maximum oxidation state is attained with oxygen extends to Mn_2O_7 [Mn(+7)] and with fluorine to CrF_6 [Cr(+6)]. In the second and third rows the maximum oxidation state range extends further, going up to M(+8) in MO_4 (M = Ru, Os) with oxygen and Re(+7) in ReF_7 with fluorine. The range over which maximum oxidation states are attained with chlorine is more limited than for fluorine, reaching only Ti(+4) in $TiCl_4$ for the first row and W(+6) in WCl_6 for the third row. The onset of a maximum oxidation state of +2 also starts earlier with chlorine than with fluorine.

In the f-block the lanthanide elements show quite different behaviour, and oxidation state +3 dominates the binary oxides and fluorides formed

Table 3.3 Simple binary transition metal chlorides[a]

Sc	Ti	V	Cr	Mn	Fe	Co	Ni	Cu	Zn
								CuCl	
	$TiCl_2$	VCl_2	$CrCl_2$	$MnCl_2$	$FeCl_2$	$CoCl_2$	$NiCl_2$	$CuCl_2$	$ZnCl_2$
$ScCl_3$	$TiCl_3$	VCl_3	$CrCl_3$	$MnCl_3$	$FeCl_3$				
	$TiCl_4$	VCl_4	$CrCl_4$						
Y	**Zr**	**Nb**	**Mo**	**Tc**	**Ru**	**Rh**	**Pd**	**Ag**	**Cd**
								AgCl	
	$ZrCl_2$	Nb_6Cl_{14}	$MoCl_2$		$RuCl_2$		$PdCl_2$		$CdCl_2$
YCl_3	$ZrCl_3$	$NbCl_3$	$MoCl_3$		$RuCl_3$	$RhCl_3$			
	$ZrCl_4$	$NbCl_4$	$MoCl_4$	$TcCl_4$					
		$NbCl_5$	$MoCl_5$						
			$MoCl_6$	$TcCl_6$					
La	**Hf**	**Ta**	**W**	**Re**	**Os**	**Ir**	**Pt**	**Au**	**Hg**
								AuCl	Hg_2Cl_2
		Ta_2Cl_5	WCl_2		$OsCl_2$		$PtCl_2$		$HgCl_2$
$LaCl_3$		$TaCl_3$	WCl_3	$ReCl_3$	$OsCl_3$	$IrCl_3$		$AuCl_3$	
	$HfCl_4$	$TaCl_4$	WCl_4	$ReCl_4$	$OsCl_4$		$PtCl_4$		
		$TaCl_5$	WCl_5	$ReCl_5$					
			WCl_6	$ReCl_6$					

[a]Except for Ta_2Cl_5 and Nb_6Cl_{14}, chlorides with non-integer oxidation states have been excluded

across the series. Only cerium, praesodymium and terbium appear with a higher oxidation state of +4 in their oxides and fluorides (Tables 3.4 and 3.5). Among the lanthanide chlorides, none appears with an oxidation state higher than +3 (Table 3.6). The behaviour of the early actinides is more similar to that of the d-block elements in that the maximum oxidation state possible is attained out to UO_3 with oxygen and UF_6 with fluorine (Tables 3.4 and 3.5). However, by the time americium is reached, oxidation state +4 has become the observed maximum, extending as far as californium in both oxides and fluorides. Among the chlorides of the actinide series, the group oxidation state is attained as far as UCl_6 but thereafter $NpCl_4$ is followed by $AnCl_3$, from An = Pu to Es.

3.2.2 Variation in Oxidation States

In addition to the maximum values attainable, the variety of oxidation states for which compounds can be isolated is of importance. Among the first-row oxides, scandium in Group 3 and zinc in Group 12 appear in only the group oxidation state. In the second and third row, zirconium

Table 3.4 Simple binary oxides of the f-block elements[a]

Ce	Pr	Nd	Pm	Sm	Eu	Gd	Tb	Dy	Ho	Er	Tm	Yb	Lu
		NdO		SmO	EuO							YbO	
Ce_2O_3	Pr_2O_3	Nd_2O_3		Sm_2O_3	Eu_2O_3	Gd_2O_3	Tb_2O_3	Dy_2O_3	Ho_2O_3	Er_2O_3	Tm_2O_3	Yb_2O_3	Lu_2O_3
CeO_2	PrO_2						TbO_2						
Th	**Pa**	**U**	**Np**	**Pu**	**Am**	**Cm**	**Bk**	**Cf**	**Es**	**Fm**	**Md**	**No**	**Lr**
				Pu_2O_3	Am_2O_3	Cm_2O_3	Bk_2O_3	Cf_2O_3	Es_2O_3				
ThO_2	PaO_2	UO_2	NpO_2	PuO_2	AmO_2	CmO_2	BkO_2	CfO_2					
	Pa_2O_5	U_2O_5	Np_2O_5										
		UO_3											

[a]Oxides with non-integer oxidation states have been excluded

Table 3.5 Simple binary fluorides of the f-block elements[a]

Ce	Pr	Nd	Pm	Sm	Eu	Gd	Tb	Dy	Ho	Er	Tm	Yb	Lu
				SmF_2	EuF_2							YbF_2	
CeF_3	PrF_3	NdF_3		SmF_3	EuF_3	GdF_3	TbF_3	DyF_3	HoF_3	ErF_3	TmF_3	YbF_3	LuF_3
CeF_4	PrF_4						TbF_4						
Th	**Pa**	**U**	**Np**	**Pu**	**Am**	**Cm**	**Bk**	**Cf**	**Es**	**Fm**	**Md**	**No**	**Lr**
		UF_3	NpF_3	PuF_3	AmF_3	CmF_3	BkF_3	CfF_3	EsF_3				
ThF_4	PaF_4	UF_4	NpF_4	PuF_4	AmF_4	CmF_4	BkF_4	CfF_4					
	PaF_5	UF_5	NpF_5										
		UF_6	NpF_6	PuF_6									

[a]Fluorides with non-integer oxidation states have been excluded

Table 3.6 Simple binary chlorides of the f-block elements[a]

Ce	Pr	Nd	Pm	Sm	Eu	Gd	Tb	Dy	Ho	Er	Tm	Yb	Lu
		$NdCl_2$		$SmCl_2$	$EuCl_2$			$DyCl_2$			$TmCl_2$	$YbCl_2$	
$CeCl_3$	$PrCl_3$	$NdCl_3$		$SmCl_3$	$EuCl_3$	$GdCl_3$	$TbCl_3$	$DyCl_3$	$HoCl_3$	$ErCl_3$	$TmCl_3$	$YbCl_3$	$LuCl_3$
Th	**Pa**	**U**	**Np**	**Pu**	**Am**	**Cm**	**Bk**	**Cf**	**Es**	**Fm**	**Md**	**No**	**Lr**
					$AmCl_2$			$CfCl_2$	$EsCl_2$				
		UCl_3	$NpCl_3$	$PuCl_3$	$AmCl_3$	$CmCl_3$	$BkCl_3$	$CfCl_3$	$EsCl_3$				
$ThCl_4$	$PaCl_4$	UCl_4	$NpCl_4$										
	$PaCl_5$	UCl_5											
		UCl_6											

[a]Chlorides with non-integer oxidation states have been excluded

and hafnium appear only in their maximum oxidation states with oxygen, while gold only forms Au_2O_3. Similar trends are apparent for the d-block metal fluorides. In the first and second rows of the d-block, variable oxidation states appear in Group 4, but for the third row, binary fluorides appear only as their group oxidation state until rhenium is reached. The onset of an oxidation state maximum of +2 in binary fluorides occurs in Group 10 for the first-row, Group 11 for the second-row and Group 12 for the third-row elements. Among the early third-row elements, a greater range of oxidation states is found with the chlorides than with the fluorides. These trends in chemical behaviour are determined by the overall thermodynamics of compound formation. In turn, these reflect the underlying electronic structures of the transition elements as expressed in their IE, EA and χ values and their radii.

3.3 Thermodynamic Aspects of Compound Formation

3.3.1 The Born–Haber Cycle

The reaction of a metal, M, with a non-metallic element, E_n, to form a binary compound, ME_x, is summarized in equation 3.1:

$$M_{(s)} + (^x/_n)E_{n(s)} \rightarrow ME_{x(s)} \qquad (3.1)$$

However, this simple chemical equation conceals a more complicated sequence of events in which the reactants undergo various transformations before the product is formed. These may be summarized in a Born–Haber thermodynamic cycle (Figure 3.1). The first stage of the reaction process is the conversion of M and E_n into gaseous state atoms, requiring an enthalpy of atomization of $M_{(s)}$, $\Delta H^\ominus_{at}(M)$, and, if E_n is a solid or liquid, the enthalpy of vaporization, $\Delta H^\ominus_{vap}(E_n)$ of $E_{n(s)}$ or $E_{n(l)}$. In addition, if n is not equal to 1, the total bond dissociation enthalpy, $\Sigma^n BDE(E_n)$, required to convert $E_{n(g)}$ into $nE_{(g)}$, will have to be included. The atoms of $M_{(g)}$ will then need to be ionized to $M^{z+}_{(g)}$, absorbing the appropriate sum of IEs, $\Sigma^z IE(M)$, in the process. Conversion of the atoms of $E_{(g)}$ into the ions $E^{(z/x)-}_{(g)}$ will involve the sum of the EAs, $\Sigma^{(z/x)}EA(xE)$. These gaseous ions may then condense to form a solid $ME_{x(s)}$, releasing a lattice enthalpy, $\Delta H^\ominus_U$. The cycle is completed by the enthalpy of formation of ME_x, $\Delta H^\ominus_f(ME_x)$. This whole process can be represented by equation 3.2:

$$\Delta H^\ominus_f(ME_x) = \Delta H^\ominus_{at}(M) + \Delta H^\ominus_{vap}(E_n) + \Sigma^n BDE(E_n) + \Sigma^z IE(M) + \Sigma^{(z/x)}EA(xE) + \Delta H^\ominus_U \qquad (3.2)$$

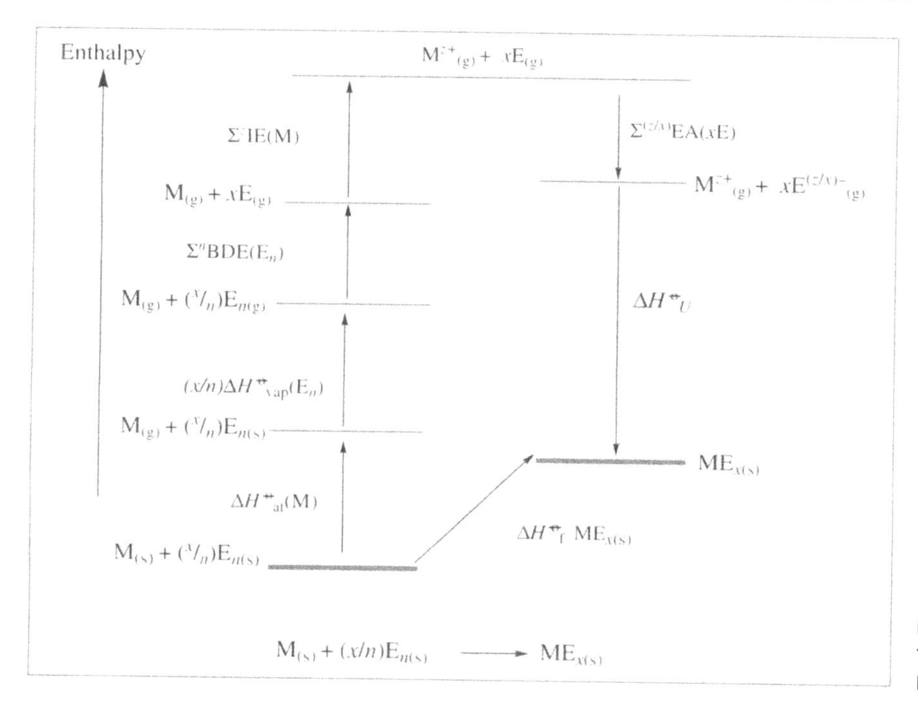

Figure 3.1 A Born–Haber cycle for the formation of an ionic compound from its elements

which equates the heat of formation of ME_x to the sum of the other parameters involved in the compound forming process (Box 3.1). This thermodynamic cycle applies to the formation of an ionic compound, and it is possible to calculate the lattice enthalpy of purely ionic compounds using the Born–Landé equation (equation 3.3) in which N_A is the Avogadro number, M the Madelung constant, z^+ and z^- are the respective charges on the cation and anion, e the electronic charge, ε_0 the permittivity of a vacuum, and r_0 the interionic distance, *i.e.* the sum of the cation and anion radii, respectively r^+ and r^-:

$$\Delta H^{\theta}_U = -\frac{N_A M z^+ z^- e^2}{4\pi\varepsilon_0 r_0}(1 - 1/n) \tag{3.3}$$

An ion in a crystal lattice lies in an electrostatic field created by all the surrounding layers of counter ions and like ions. This field will depend upon the exact geometric structure of the lattice. It is this which determines the various distances between the ion and other counter ions or like ions, as well as the numbers of each type at particular distances. The **Madelung constant** takes account of this geometric arrangement and, in effect, represents the total effect of the various layers of counter ions and like ions on the electrostatic field experienced by an ion. Values of the Madelung constant have been calculated for the various types of crystal lattice which can form.

Box 3.1 Born–Haber Cycle Calculations

An example of a Born–Haber cycle calculation is provided by the formation of $TiCl_{2(s)}$ from titanium and chlorine, as shown in Figure 3.2.

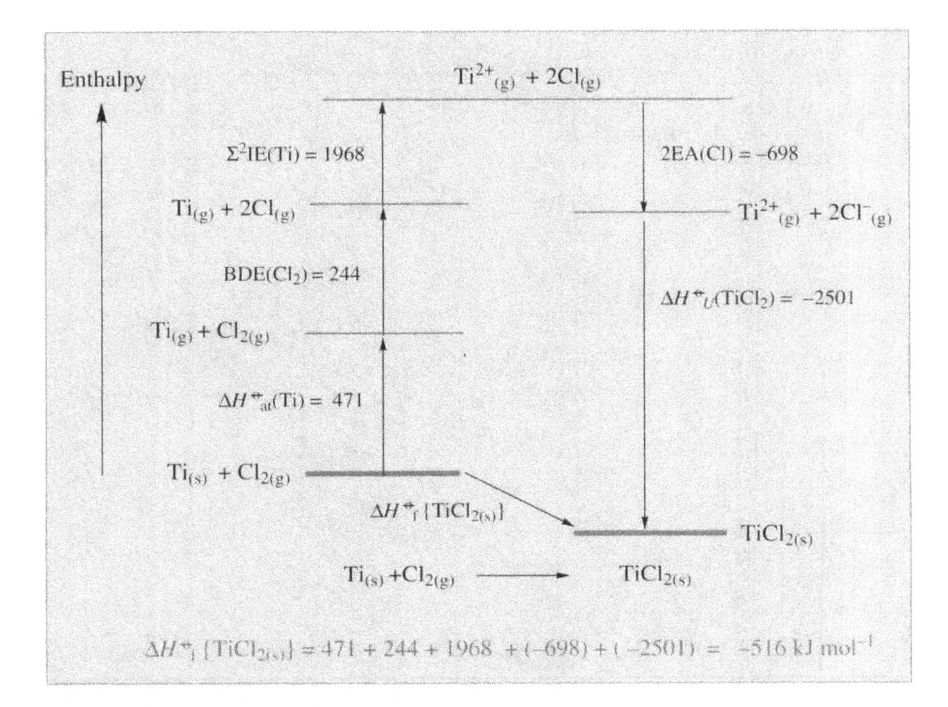

Figure 3.2 A Born–Haber cycle for the formation of $TiCl_2$ from its elements

Worked Problem 3.1

Q Using a Born–Haber Cycle, calculate the enthalpy of formation of $TiCl_{4(l)}$ from titanium and chlorine. Use the data from Box 3.1, 6.828 MJ mol^{-1} for the sum of the 3rd and 4th IEs of titanium and a value of −9431 kJ mol^{-1} for the enthalpy of forming liquid $TiCl_4$ from its component gaseous ions.

A Following the example given in Box 3.1 and using the data therein, a similar calculation may be performed for $TiCl_{4(l)}$:

$$Ti_{(s)} + 2Cl_{2(g)} \rightarrow TiCl_{4(l)}$$

$$\Delta H^{\ominus}_{f}\{TiCl_{4(l)}\} = 471 + 488 + 8796 - 1396 - 9431 = -1072 \text{ kJ mol}^{-1}$$

The variable n is known as the Born exponent, which depends upon the electronic configuration of the ions present. The Born–Landé equation calculates $\Delta H^{\ominus}_{U}$ on the basis of the electrostatic attraction between the ions in a lattice of known structure and, for alkali metal halides, typically predicts experimental values to within a few percent. The calculated and experimental values of $\Delta H^{\ominus}_{U}$ for NaCl, for example, are −763 and –778 kJ mol^{-1}, respectively, a difference of only 2%. However, this agreement between theory and experiment is less typical for transition metal compounds. In the case of AgCl the calculated and experimental values of $\Delta H^{\ominus}_U$ are –734 and –905 kJ mol^{-1}, respectively, a difference of 19%. This indicates that the bonding in AgCl cannot be viewed as purely ionic and that some covalent contribution to $\Delta H^{\ominus}_U$ is present. In this case the covalent contribution is modest compared to the ionic bonding, but in other cases the covalent interaction is sufficiently important for molecular compounds to form, examples being $TiCl_4$ and VCl_4 which are liquids at room temperature. This behaviour may be anticipated from the small differences in electronegativity between certain transition metals and chlorine compared to the alkali metals and chlorine.

Worked Problem 3.2

Q Using the Born–Landé equation, calculate the lattice energy of TiO_2, which has the rutile structure with a Madelung constant, M, of 2.408. The effective ionic radii of Ti^{4+} ($z^+ = 4$, CN = 6) and O^{2-} ($z^- = 2$, CN = 3) are respectively 60.5 and 136 pm, with a Born exponent, n, of 8. The Avogadro number, N_A is 6.022×10^{23} mol^{-1}, the permittivity of a vacuum, ε_0, is 8.854×10^{-12} F m^{-1} and the electron charge, e, is 1.602×10^{-19} C.

A Incorporating the values given into equation 3.3 leads to a calculated lattice energy for TiO_2 as follows:

$$\Delta H^{\ominus}_U = -\frac{6.022\times10^{23}\times2.408\times4\times2\times(1.602\times10^{-19})^2}{4\times3.142\times8.854\times10^{-12}\times(60.5+136)\times10^{-12}}\times(1-1/8)$$

$$= -\frac{297.7\times10^{-15}}{21866\times10^{-24}}\times0.875 = -11913 \text{ kJ mol}^{-1}$$

Whether or not a compound ME_x is stable at a particular temperature will depend upon whether decomposition reactions, exemplified by equations 3.4 and 3.5, or disproportionation, exemplified by equation 3.6, are thermodynamically favourable.

$$ME_x \rightarrow ME_{(x-1)} + E \quad (3.4)$$

$$xME \rightarrow ME_x + (x-1)M \quad (3.5)$$

$$2ME_x \rightarrow ME_{(x-1)} + ME_{(x+1)} \quad (3.6)$$

The vanadium halides provide an example. The liquid tetrachloride VCl_4 can be synthesized from vanadium and excess elemental chlorine at 500 °C. However, on heating in the absence of chlorine, VCl_4 loses chlorine to form VCl_3, which in turn loses chlorine on further heating to give VCl_2, which is stable at its melting point of 1350 °C (equations 3.7–3.9).

$$V_{(s)} + 2Cl_{2(g)} \rightarrow VCl_{4(l)} \quad (3.7)$$

$$VCl_{4(l)} \rightarrow VCl_{3(s)} + {}^1/_2Cl_{2(g)} \quad (3.8)$$

$$VCl_{3(s)} \rightarrow VCl_{2(s)} + {}^1/_2Cl_{2(g)} \quad (3.9)$$

Disproportionation reactions may also occur, as exemplified by equations 3.10 and 3.11 (see Box 3.2):

$$2VCl_{3(s)} \rightarrow VCl_{2(s)} + VCl_{4(l)} \quad (3.10)$$

$$2VF_{4(s)} \rightarrow VF_{3(s)} + VF_{5(l)} \quad (3.11)$$

Similarly, $CrCl_3$ can be sublimed at about 600 °C in a stream of chlorine, but in the absence of chlorine decomposes to $CrCl_2$.

Box 3.2 Decomposition and Disproportionation

The standard free energy (ΔG^{Θ}) of the reaction in which VCl_4 loses chlorine to give VCl_3, according to equation 3.8, can be calculated by subtracting the standard free energy of forming $VCl_{4(l)}$ (–504 kJ mol^{-1}) from the sum of the standard free energies of formation of $VCl_{3(s)}$ and ${}^1/_2Cl_{2(g)}$ (–511 and 0 kJ mol^{-1}, respectively). This gives a value of –7 kJ mol^{-1}, showing that the reaction is spontaneous under standard conditions, *i.e.* 25 °C and 1 bar (750 Torr) partial pressure of chlorine.

At 25 °C the decomposition of $VCl_{3(s)}$ according to equation 3.9 has a positive ΔG^{Θ} and is not spontaneous. However, the temperature, T^d (in K), at which ΔG^{Θ} changes from positive to negative (*i.e.* passes through zero) can be estimated from the relationship between ΔG^{Θ}, the standard enthalpy change (ΔH^{Θ}), the standard entropy change (ΔS^{Θ}) and temperature (T) shown in equation 3.12. Above T^d, ΔG^{Θ} will become negative and the reaction will be spontaneous.

$$\Delta G^{\Theta} = 0 = \Delta H^{\Theta} - T^d\Delta S, \text{ so that } T^d = \Delta H^{\Theta}/\Delta S^{\Theta} \quad (3.12)$$

The standard enthalpies of formation of $VCl_{3(s)}$ and $VCl_{2(s)}$ are respectively −581 and −452 kJ mol^{-1}, and the standard entropies, $S^{\ominus}$, of these compounds are respectively +131 and +97 J K^{-1} mol^{-1}. Elemental chlorine, $Cl_{2(g)}$, has $\Delta H^{\ominus}$ = 0 kJ mol^{-1} and $S^{\ominus}$= 222 J K^{-1} mol^{-1}. Thus the values of $\Delta H^{\ominus}$ and $\Delta S^{\ominus}$ for reaction 3.9 can be calculated by subtracting the values for the reactants from the values for products as follows:

$$\Delta H^{\ominus} = -452 + 0 - (-581) = +129 \text{ kJ mol}^{-1}$$
$$\Delta S^{\ominus} = 97 + {}^{1}/_{2}(222) - 131 = +77 \text{ J K}^{-1} \text{ mol}^{-1}$$

These values can be used to determine the value of T^{d} as follows:

$$\Delta H^{\ominus}/\Delta S^{\ominus} = T^{d} = 129000/77 = 1675 \text{ K or } 1402 \text{ °C}$$

Such calculations assume that the entropy and enthalpy terms are independent of temperature. Although this assumption is usually reasonable over small temperature ranges, it may not always be valid when T^{d} is very different from the standard condition of 25 °C. Thus the results of the above calculations can only provide estimates of T^{d}.

Worked Problem 3.3

Q Calculate the temperature, T^{d}, at which the free energy for the decomposition of $VCl_{2(s)}$ into $V_{(s)}$ and $Cl_{2(g)}$ according to equation 3.13 becomes zero. The standard enthalpy for $V_{(s)}$ is 0 kJ mol^{-1} and the standard entropy is 29 J K^{-1} mol^{-1}. Values for chlorine are given in Box 3.2.

$$VCl_{2(s)} \rightarrow V_{(s)} + Cl_{2(g)} \quad (3.13)$$

A A similar calculation to that in Box 3.2 leads to a T^{d} value of 2662 °C for the decomposition of $VCl_{2(s)}$ in accord with the thermal stability of this compound:

$$VCl_{2(s)} \rightarrow V_{(s)} + Cl_{2(g)}$$
$$\Delta H^{\ominus} = 0 + 0 - (-452) = +452 \text{ kJ mol}^{-1}$$
$$\Delta S^{\ominus} = 29 + 222 - 97 = +154 \text{ J K}^{-1} \text{ mol}^{-1}$$
$$\Delta H^{\ominus}/\Delta S^{\ominus} = T^{d} = 452000/154 = 2935 \text{ K or } 2662 \text{ °C}$$

3.3.2 Contributions from Covalency

The contribution to bonding from covalency is important in the formation of higher oxidation state compounds, since ionic bonding alone could not provide the energy needed to form the high oxidation state ions. The Born–Landé equation reveals how lattice energies for ionic bonding vary with the product of the cation and anion ionic charges and inversely with the distance between ions in the lattice. Thus a higher oxidation state ion with a larger z^+ and a smaller r^+ should give rise to a larger lattice energy. There will also be an increased contribution from the additional EAs involved, but these effects must be set against the additional BDEs which may be involved and the rapidly increasing magnitude of the higher IE values. As the oxidation state increases, the price in increased total ionization enthalpy cannot be met from a purely ionic bonding model, and covalent interactions must contribute. In OsO_4, for example, although the formal oxidation state of osmium is +8, it is not reasonable to assume that the molecule contains Os^{8+}, just as we would not normally describe CO_2 as containing C^{4+}. As the cationic charge increases and the ionic radius decreases, the ion becomes more and more polarizing and so more and more able to distort the electron cloud round the counter ion so that the appropriate bonding model becomes less ionic and more covalent. This effect is enhanced by the presence of larger, more highly charged anions which are more polarizable.

The effects of an absence of significant covalency in bonding can be seen in the halides and oxides of the lanthanides compared to those of the d-block elements (Tables 3.1–3.6). In the lanthanide Ln^{3+} ions the 5d orbitals are empty and electrons in the core-like 4f orbitals are unable to enter into covalent bonding to any significant extent. Thus, although the sum of the first, second and third IEs is lower for the lanthanides than for the early d-block elements, oxidation state +3 is the maximum which can normally be attained, because the bonding is predominantly ionic in character. The early actinide elements show variable oxidation state behaviour rather like the early d-block elements, despite the sums of their first, second and third IEs being very similar to those of the lanthanides. In the early actinides, relativistic effects lead to the 6d and 5f orbitals being similar in energy and less core-like than the corresponding lanthanide valence shell orbitals. This allows some covalent contribution to bonding and hence higher oxidation states become accessible. By the middle of the actinide series the 6d and 5f orbitals are sufficiently contracted by the increasing Z_{eff} that both start to behave more as core electron shells, and the behaviour of the elements becomes like that of the lanthanide series, with oxidation state +3 becoming the maximum at einsteinium.

Since the radii of d-block ions decrease with increasing atomic number across the rows, it might be expected from the Born–Landé equation that lattice enthalpies might increase with Z. A plot of lattice enthalpy against Z for the first-row d-block metal dichlorides shows that this is the case, but that the trend is not uniform (Figure 3.3). The values for $CaCl_2$, $MnCl_2$ and $ZnCl_2$ lie on a curve, but the values for the other elements fall below this curve. Their lattice enthalpies are larger than a simple ionic model would suggest. A uniform covalent contribution to bonding would not be expected to produce the observed 'double dip', and the origins of this feature lie in the effect which the ions surrounding the metal ion have on the energies of the different d orbitals. This can be explained using the crystal field theory model presented in Chapter 6.

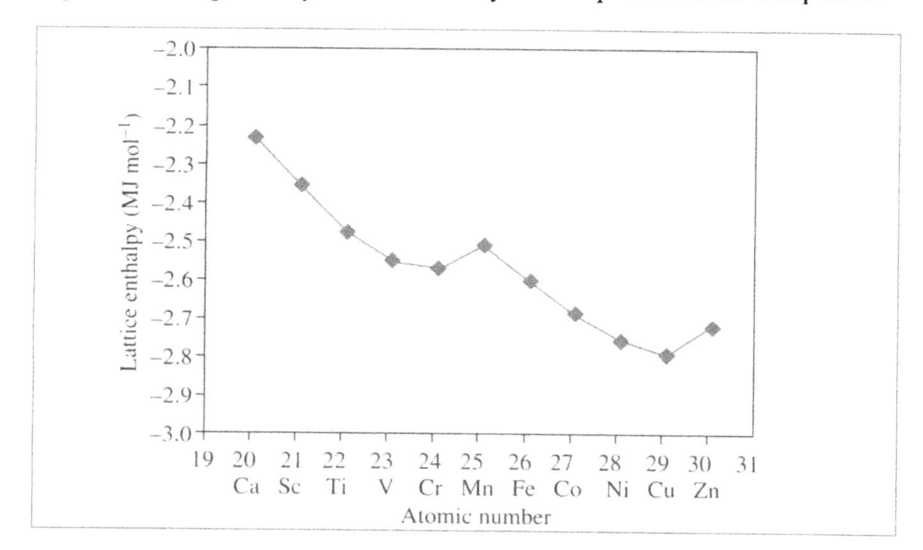

Figure 3.3 The variation in MCl_2 lattice enthalpy with atomic number across the first row of the d-block

Summary of Key Points

1. *The stoichiometries of the binary oxides, fluorides and chlorides* formed by the transitional elements provide an insight into the range of oxidation states which are chemically accessible for each element.

2. *The thermodynamics of binary compound formation* reflect the atomic and ionic properties of the metal and may be used to predict the stablility or otherwise of simple compounds.

3. *Lattice enthalpies* are important thermodynamic parameters and depend upon the solid state structures of the compound and hence the ionic radius of the metal ion.

4. *Covalency in bonding* leads to deviations from calculated lattice enthalpies based on ionic structures.

Problems

3.1. A plot of atomization enthalpy against atomic number for the lanthanides is shown in Figure 3.4. Refer to Figure 2.5 and account for the appearance of the maxima in this plot. (A consideration of the solution to Worked Problem 2.4 may help in finding an answer).

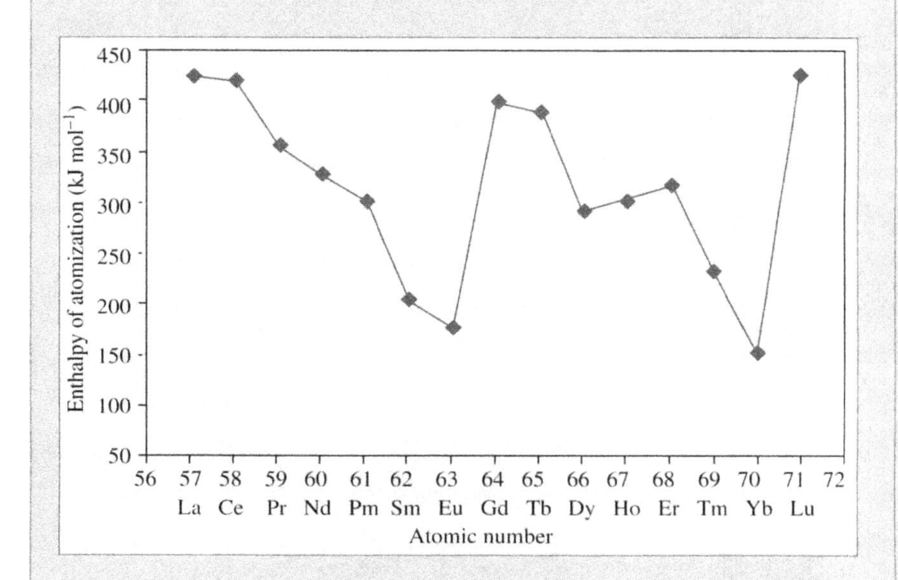

Figure 3.4 The variation in atomization enthalpy with atomic number for the lanthanides

3.2. Using the Born–Landé equation and the data below, calculate the lattice enthalpies of MnO and MnO_2.

$\varepsilon_0 = 8.854 \times 10^{-12}$ F m^{-1}; $N_A = 6.022 \times 10^{23}$ mol^{-1}; $e = 1.602 \times 10^{-19}$ C; $n = 8$.

MnO has the rock salt structure with $M = 1.748$; effective ionic radius Mn^{2+} (CN 6) = 83 pm; effective ionic radius O^{2-} (CN 6) = 140 pm.

MnO_2 has the rutile structure with $M = 2.408$; effective ionic radius Mn^{4+} (CN 6) = 53 pm; effective ionic radius O^{2-} (CN 3) = 136 pm.

3.3. Using Born–Haber cycles, the solutions to Problem 3.2 and the data below, calculate the enthalpies of formation of MnO and MnO_2.

IE values for Mn: 1st = +717, 2nd = +1509, 3rd = +3248, 4th = +4940 kJ mol^{-1}.
ΔH^{o}_{at}(Mn) = +220 kJ mol^{-1}; BDE(O_2) = +498 kJ mol^{-1}
EA(O) = −141; EA(O^-) = +844 kJ mol^{-1}.

3.4. Determine the temperature at which the disproportionation reaction of $VCl_{3(s)}$, shown in equation 3.10, becomes spontaneous. Use the thermodynamic data provided in Box 3.2 for $VCl_{2(s)}$ and $VCl_{3(s)}$, and the values for $VCl_{4(l)}$: ΔH^{o} = −569 kJ mol^{-1} and ΔS^{o} = +255 J K^{-1} mol^{-1}.

3.5. Discuss the thermal stability of the chlorides of uranium based on the thermodynamic data provided in Table 3.7.

Table 3.7

Compound	UCl_3	UCl_4	U_2Cl_{10}	UCl_6	Cl_2	*Units*
ΔH^{o}	−891	−1050	−2192	−1138	0	kJ mol^{-1}
S^{o}	+165	+197	+470	+285	+222	J K^{-1} mol^{-1}

3.6. On the basis of the data in Table 3.8, can MnO_2 oxidize V_2O_4 to V_2O_5 at 800 °C, assuming that only MnO is formed from MnO_2?

Table 3.8

Compound	MnO_2	MnO	V_2O_5	V_2O_4	*Units*
ΔH^{o}	−521	−385	−1560	−1439	kJ mol^{-1}
S^{o}	+53	+60	+131	+103	J K^{-1} mol^{-1}

4
Coordination Compounds

Aims

By the end of this chapter you should understand the terms:

- Coordination compound
- Coordination number
- Coordination geometry
- Uni-, bi- and poly-dentate as applied to ligands
- Homo- and hetero-leptic as applied to complexes

and have a knowledge of:

- The structures of the regular coordination polyhedra of coordination compounds
- The types of isomerism possible in transition element complexes
- Some representative types of ligand which may appear in transition element complexes

When considering the **structures** of coordination compounds it is worth noting that transition element complexes are usually formed from reactions between their salts and Brønsted bases in solution. However, the structures of the compounds formed are usually determined in the solid state using samples crystallized *from solution*. While it may usually be assumed that the solid state structures are similar to the solution structures, this may not always be so, and some complexes may adopt different structures in solution and the solid state.

4.1 Introduction

In order to explain the formulae and structures of the complex compounds, or complexes, formed by transition metal salts with molecular species such as ammonia, Werner coined the terms *primary valence* and *secondary valence*, as explained in Chapter 1. These concepts remain valid today except that the term oxidation state has replaced 'primary valence' and the term coordination number has replaced 'secondary valence'. Werner had recognized that a transition metal salt could form a complex compound in which the metal ion became bonded to a number of groups which need not necessarily be the counter anions originally present in the salt. The orientations in space of these metal-bound groups would lead to the complex having a particular geometric structure. In this chapter the structures of transition element complexes are examined in more detail and some definitions of key terms are provided.

4.2 Coordination Compounds

4.2.1 Complexes

One definition of a metal complex or coordination compound is '*a compound formed from a Lewis acid and a Brønsted base*', a Lewis acid being an electron pair acceptor and a Brønsted base a proton acceptor. Thus the interaction of the Lewis acid metal centre in $Ni(ClO_4)_2$ with the Brønsted base ammonia to form a *complex* according to equation 4.1

$$Ni(ClO_4)_2 + 6NH_3 \rightarrow [Ni(NH_3)_6](ClO_4)_2 \quad (4.1)$$

provides an example of the formation of a coordination compound. In writing the formulae of metal complexes it is conventional to include the complete coordination complex within square brackets, an example being provided by $[Co(NH_3)_5Cl]Cl_2$, in which the coordination complex is $[Co(NH_3)_5Cl]^{2+}$ with two chloride counterions. The Brønsted bases attached to the metal ion in such compounds are called ligands. These may be simple ions such as Cl^-, small molecules such as H_2O or NH_3, larger molecules such as $H_2NCH_2CH_2NH_2$ or $N(CH_2CH_2NH_2)_3$, or even macromolecules, such as proteins.

Strictly speaking, the term 'ligand' only applies to groups attached to a metal ion. The term **proligand** may be used to refer to a species which may become a ligand through being bound to a metal ion but is not presently in a complex. In this text the symbols X, Y, Z will be used to represent ligands such as F, Cl, NH_3 or H_2O in which one donor atom is bound to the metal ion. The symbol L-L will be used to denote a ligand such as $H_2NCH_2CH_2NH_2$ which is bound to a metal ion through two donor atoms. The general symbol L will be used to represent a ligand of any type.

The coordination number (CN) of a metal ion in a complex can be defined as *the number of ligand donor atoms to which the metal is directly bonded.* In the case of $[Co(NH_3)_5Cl]^{2+}$ this will be 6, the sum of one chloride and five ammonia ligands each donating an electron pair. Although this definition usually works well for coordination compounds, it is not always appropriate for organometallic compounds. An alternative definition of CN would be *the number of electron pairs arising from the ligand donor atoms to which the metal is directly bonded.* To apply this definition, it is necessary to assume an ionic formulation and a particular oxidation state for the metal ion, so that charges can be assigned to the ligands as appropriate and the number of electron pairs determined.

4.2.2 Types of Ligand

Where a ligand is bound to a metal ion through a single donor atom, as with Cl^-, H_2O or NH_3, the ligand is said to be unidentate (the ligand binds to the metal through a single point of attachment as if it had one tooth). Where two donor atoms can be used to bind to a metal ion, as with $H_2NCH_2CH_2NH_2$, the ligand is said to be bidentate, and where several donor atoms are present in a single ligand as with $N(CH_2CH_2NH_2)_3$, the ligand is said to be polydentate. When a bi- or polydentate ligand uses two or more donor atoms to bind to a single metal ion, it is said to form a chelate complex (from the Greek for claw). Such complexes tend to be more stable than similar complexes containing unidentate ligands for reasons which will be explored in Chapter 5. A huge variety of ligands appear in coordination complexes and, to illustrate this point, some examples of common types of ligand are shown in Figures 4.1–4.4. Any of a variety of elements may function as donor atoms towards metal ions, but the most commonly encountered are probably nitrogen, phosphorus, oxygen, sulfur and the halides. In addition, a large number of compounds are known which contain carbon donor atoms; these are known as *organometallic* compounds (see page 11). Bidentate ligands may be classified according to the number of atoms in the ligand which separate the donor atoms (Figure 4.1) and hence the size of the chelate ring formed with the metal ion. Thus 1,1-ligands form a four-membered chelate ring when bound to a metal ion, 1,2-ligands a five membered ring, and so on. Cyclic compounds which contain donor atoms oriented so that they can bind to a metal ion and which are large enough to encircle it are known as macrocyclic proligands and some examples are shown in Figure 4.4. Bicyclic proligands are also known which can completely encapsulate a metal ion. Some of these systems have given the names cryptand or sepulchrate, which reflect their ability to wrap up and entomb the metal ion (Figure 4.4).

A **macrocyclic ligand** may be defined as a cyclic compound comprising a ring of at least nine atoms including at least three donor atoms oriented so as to bind to a metal ion. In this minimum form a macrocyclic ligand would occupy three adjacent coordination sites on one side of a metal ion. However, larger rings such as cyclam or porphine derivatives (Figure 4.4) may have a central cavity large enough for the metal ion to fit into the plane of the macrocycle.

Sometimes ligands can bind to more than one metal ion in a bridging arrangement, for example in $[W_2Cl_9]^{3-}$ illustrated in Figure 4.5. Certain polydentate ligands are particularly good at linking together several metal ions and are refered to as polynucleating ligands

4.2.3 Structure and Isomerism

In coordination compounds, or complexes, the transition elements may show CNs ranging from 1 up to 12 or more for some f-block elements. As there are more ways than one of arranging two, or more, ligand donor atoms about a central metal ion, structural isomers are possible. However, in practice, certain structural arrangements are more favourable energetically than others, limiting the range of commonly found structural types. The idealized regular coordination geometries of complexes with CNs from 1 to 6 are summarized in Figure 4.6, CNs 7 and 8 in Figure 4.7 and the higher CNs 9–12 in Figure 4.8. Transition metal complexes usually conform to idealized geometries in an approximate sense, although some distortions from ideality are often present.

The term **coordination geometry** refers to the structural arrangement of ligand donor atoms around a metal atom in a complex. Thus, in a four-coordinate metal complex the donor atoms could be arranged at the vertices of a tetrahedron giving a *tetrahedral coordination geometry*, or they could lie in the same plane to give a *square planar coordination geometry* (see margin note on isomers, page 10).

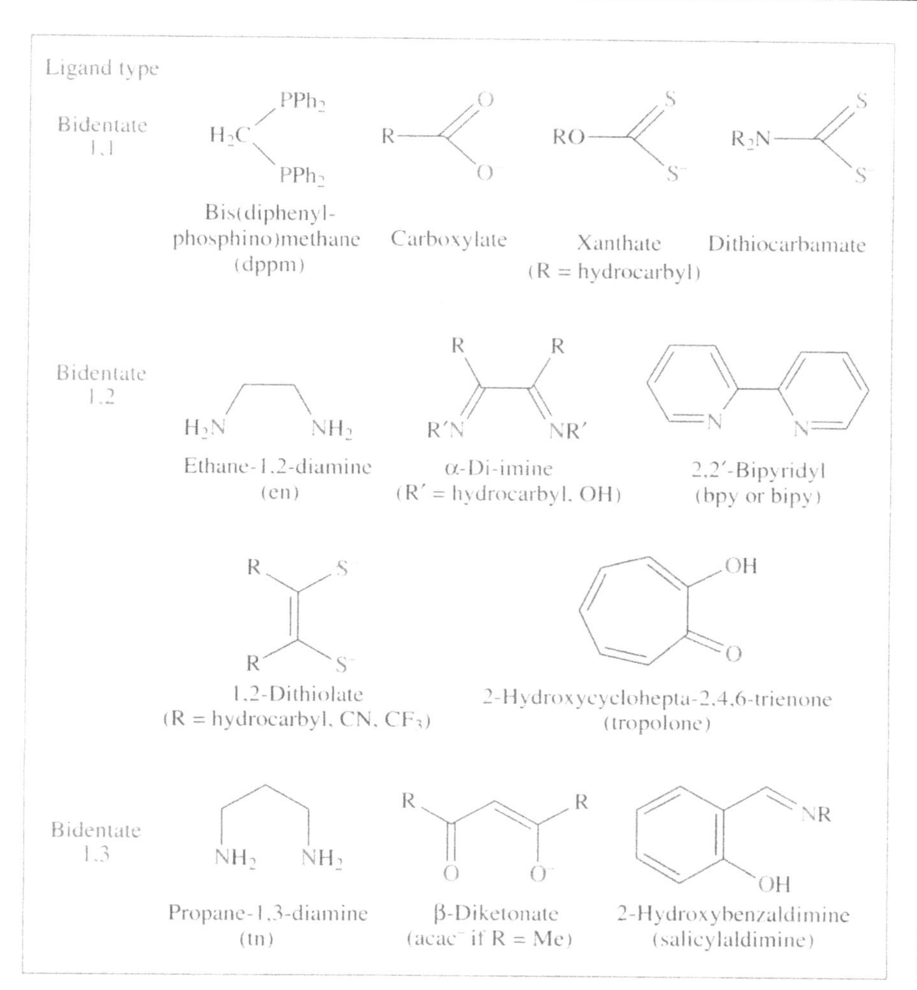

Figure 4.1 Some examples of bidentate ligands

However, the structures of the lanthanide ions tend to be less regular, particularly when more than one type of ligand is present in a heteroleptic complex.

Complexes in which a metal is bound to only one kind of donor group, e.g. $[Co(NH_3)_6]^{3+}$, are known as **homoleptic**. Complexes in which a metal is bound to more than one kind of donor group, e.g. $[Co(NH_3)_4Cl_2]^+$, are known as **heteroleptic**.

The lower CNs are rare among the transition elements. A CN of 1 is very unusual, although not unknown. Its formation depends upon the presence of a very bulky ligand which prevents the binding of additional ligands. A CN of 2 is found in some complexes of d^{10} ions such as Ag^+ or Au^+, and the geometries of these complexes are normally linear, not bent. A CN of 3 can arise in complexes with sterically demanding ligands such as the bulky amide ligand $N(SiMe_3)_2$. The d-block metal complexes $[M\{N(SiMe_3)_2\}_3]$ (M = Fe, Cr) have a trigonal planar coordination geometry rather than the T-shaped or pyramidal structures encountered with p-block elements. The f-block metal complex $[La\{N(SiMe_3)_2\}_3]$ is also thought to have a planar stucture in solution

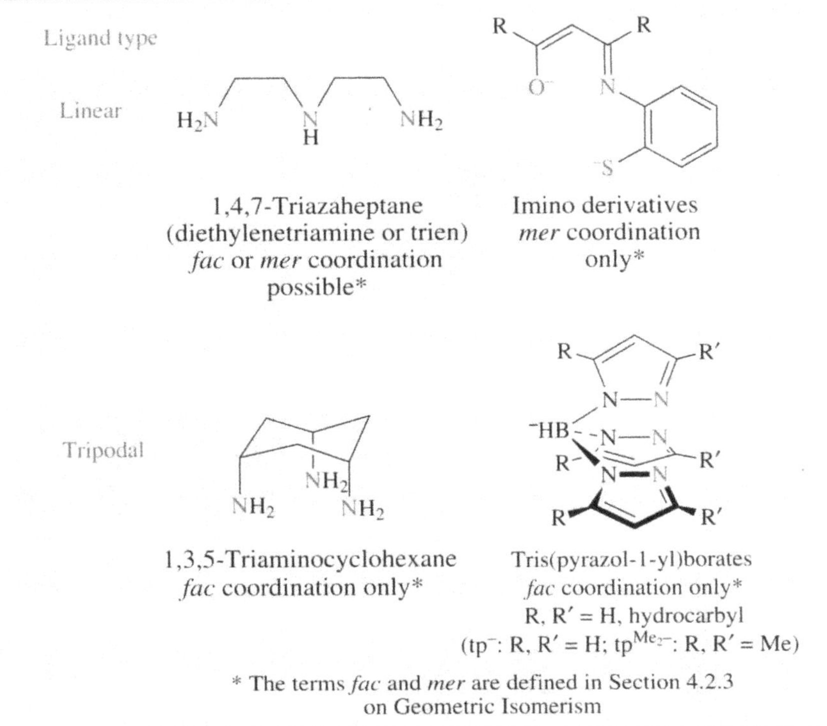

Figure 4.2 Some examples of tridentate ligands

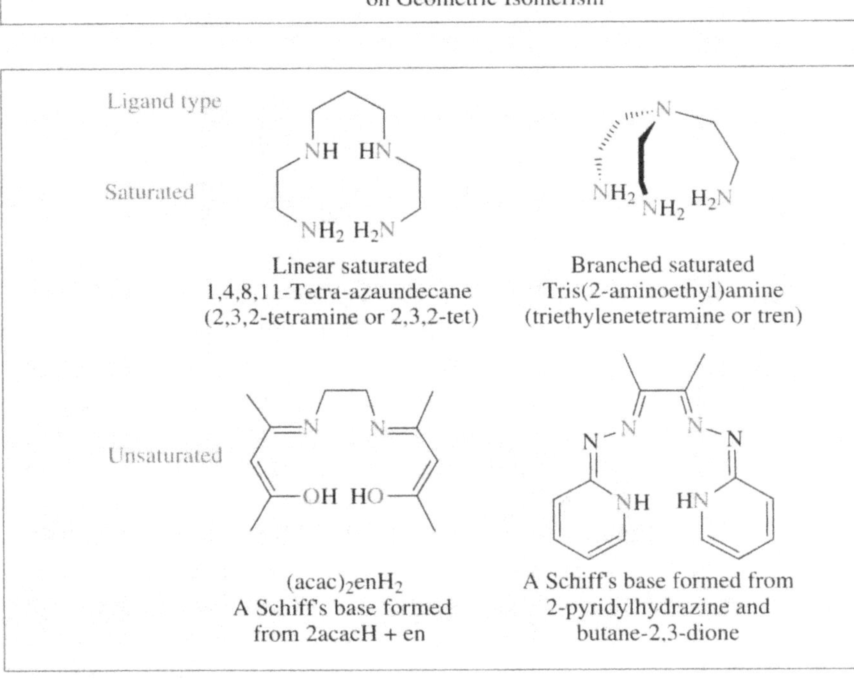

Figure 4.3 Some examples of quadridentate ligands

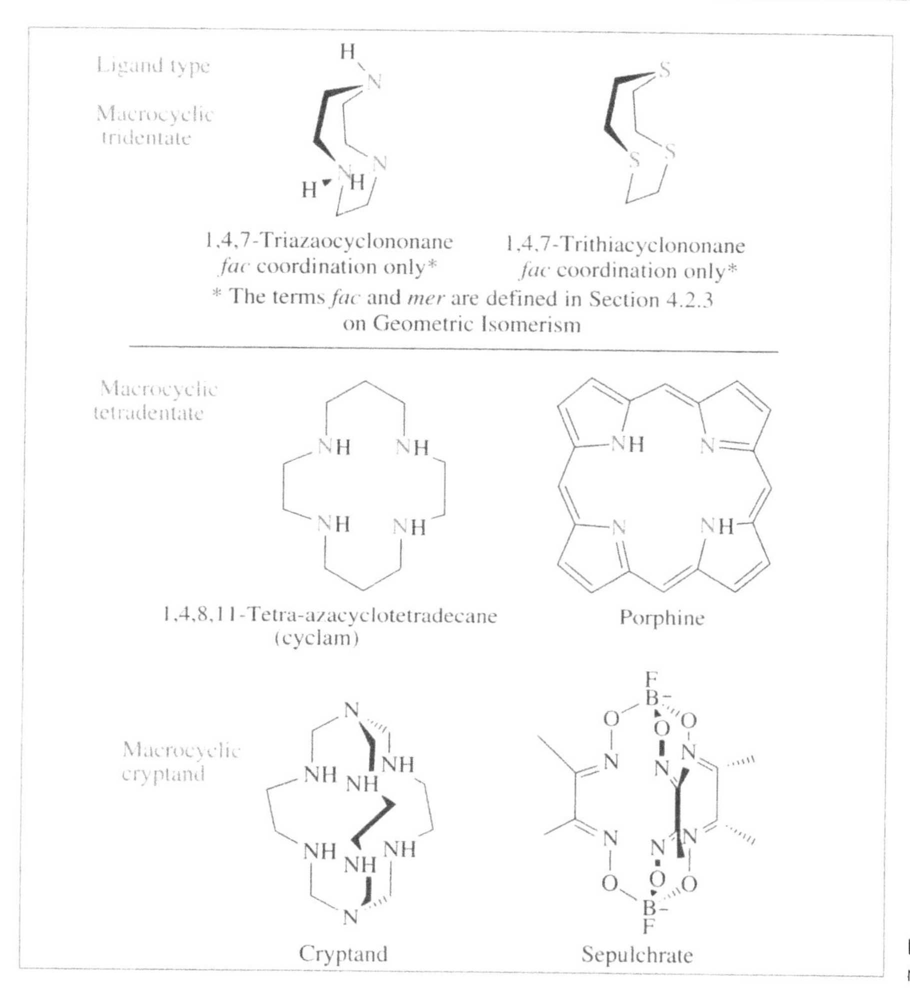

Figure 4.4 Some examples of macrocyclic ligands

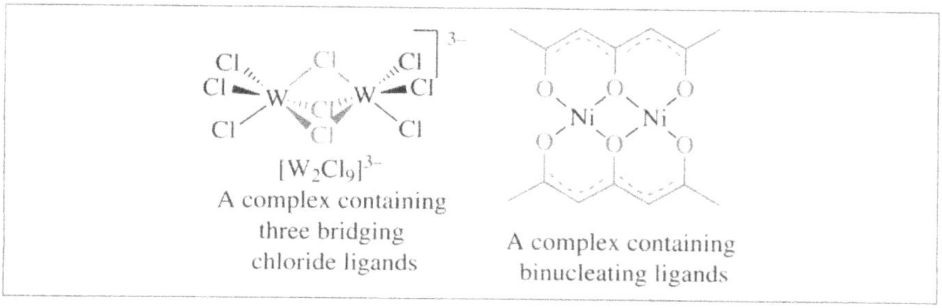

$[W_2Cl_9]^{3-}$
A complex containing three bridging chloride ligands

A complex containing binucleating ligands

Figure 4.5 Examples of a bridging ligand and a polynucleating ligand

but is pyramidal in the solid state, and provides an extremely rare example of an f-block element with a very low CN. Complexes with CN of 4 are fairly common for certain d-block metals. These complexes may be square planar or tetrahedral in geometry and, for reasons which will be explained in Chapter 6, metal ions with d^8 electron configurations tend

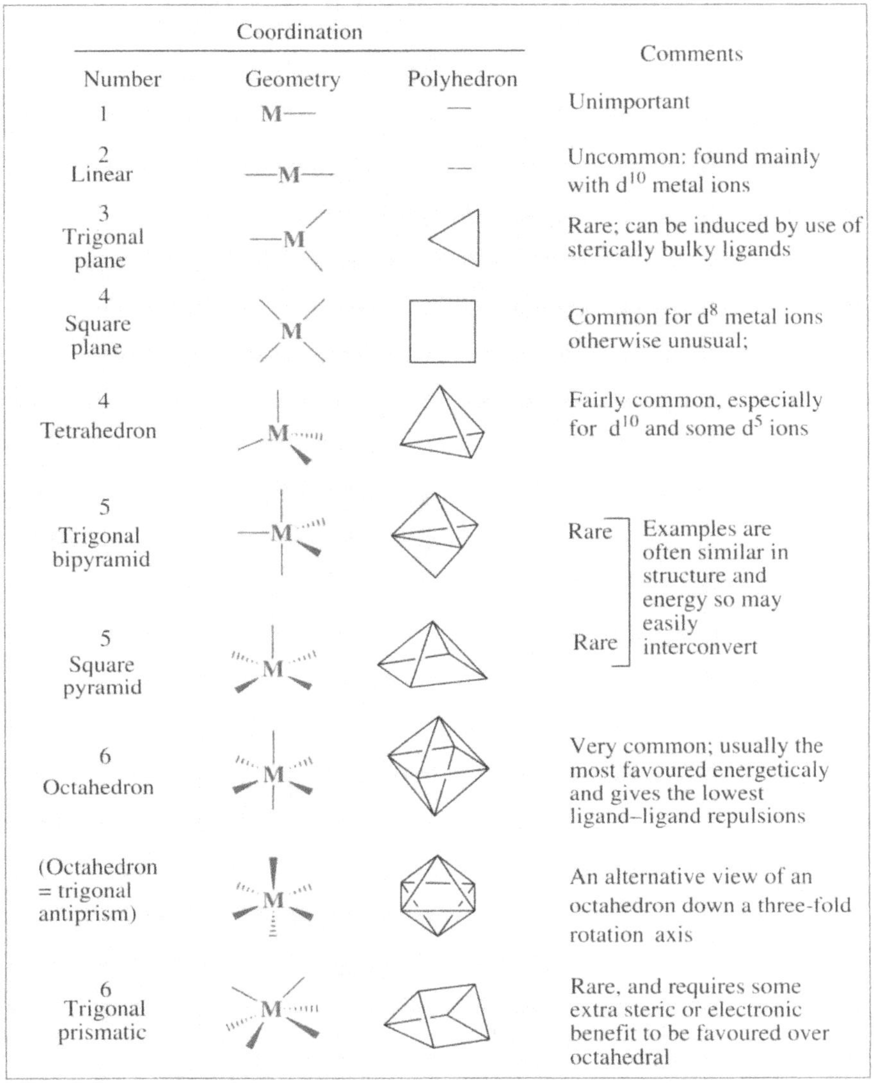

Coordination			Comments
Number	Geometry	Polyhedron	
1	M—	—	Unimportant
2 Linear	—M—	—	Uncommon: found mainly with d^{10} metal ions
3 Trigonal plane	M		Rare; can be induced by use of sterically bulky ligands
4 Square plane	M		Common for d^8 metal ions otherwise unusual;
4 Tetrahedron	M		Fairly common, especially for d^{10} and some d^5 ions
5 Trigonal bipyramid	M		Rare — Examples are often similar in structure and energy so may easily interconvert
5 Square pyramid	M		Rare
6 Octahedron	M		Very common; usually the most favoured energeticaly and gives the lowest ligand–ligand repulsions
(Octahedron = trigonal antiprism)	M		An alternative view of an octahedron down a three-fold rotation axis
6 Trigonal prismatic	M		Rare, and requires some extra steric or electronic benefit to be favoured over octahedral

Figure 4.6 Idealized structures for coordination numbers 1–6

to be square planar and those with d^5 or d^{10} configurations tetrahedral. A CN of 5 is unusual among transition element complexes and, in a complex which is purely ionically bonded, would be unstable with respect to disproportionation into CN 4 and 6 species. However, covalent contributions to bonding can stabilize CN 5. In the absence of structural demands imposed by the ligands, the two regular structures, trigonal bipyramidal and square pyramidal, are easily interconverted and similar in energy. In the solid state, both structures can be found in slightly distorted form in salts of $[Ni(CN)_5]^{3-}$.

Coordination Number	Coordination Geometry	Coordination Polyhedron	Comments
7 Pentagonal bipyramid	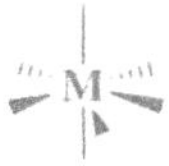 M	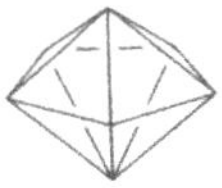	Uncommon
7 Monocapped octahedron	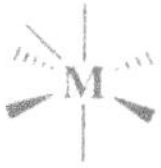 M	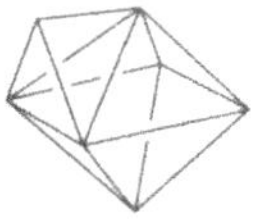	Uncommon
8 Dodecahedron	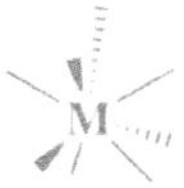 M	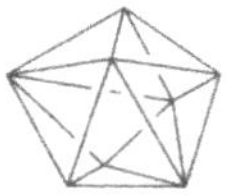	Most sterically efficient geometric arrangement for eight equivalent ligands
8 Square antiprism	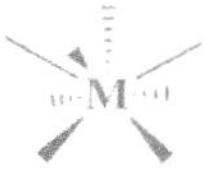 M	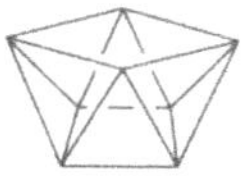	Uncommon
8 Cube	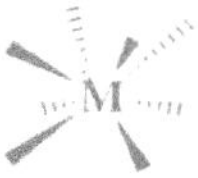 M	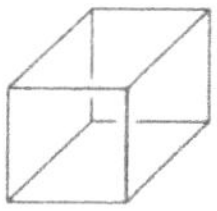	Rare; found only with the largest metal ions
8 Hexagonal bipyramid	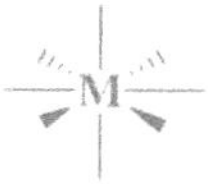 M	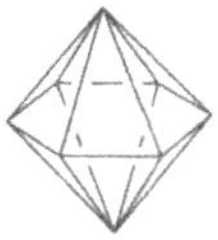	Quite common for eight-coordinate complexes of metals with *trans*-dioxo ligands

Figure 4.7 Idealized structures for coordination numbers 7 and 8

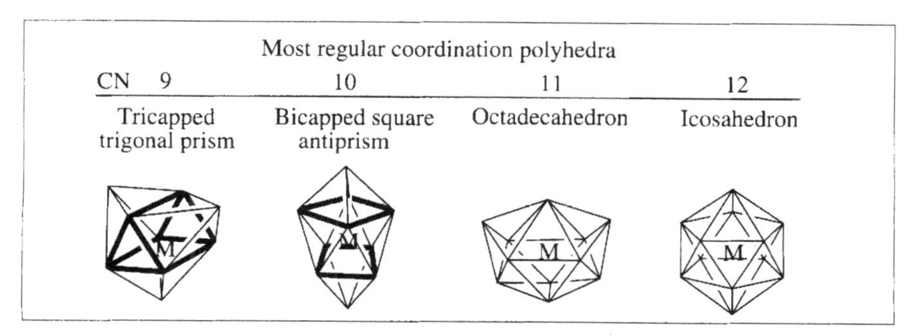

Figure 4.8 Idealized structures for coordination numbers larger than 8

The most common CN for d-block transition metal ions is 6, usually with an octahedral geometry. Ligand–ligand interactions make trigonal prismatic structures less energetically favourable so that this geometry is rare, although some examples are known, *e.g.* $[WMe_6]$ and $[Re(S_2C_2Ph_2)_3]$.

An important distortion of the octahedral structure, found in certain complexes, results from stretching or compressing the octahedron along a fourfold rotation symmetry axis, producing a tetragonal distortion (Figure 4.9a). This type of distortion is commonly found among complexes of the d^9 Cu^{2+} ion for reasons which will be explored in Chapter 6. Another, less important, type of distortion results from stretching or compressing the octahedron along a threefold rotation symmetry axis, producing a trigonal distortion (Figure 4.9b).

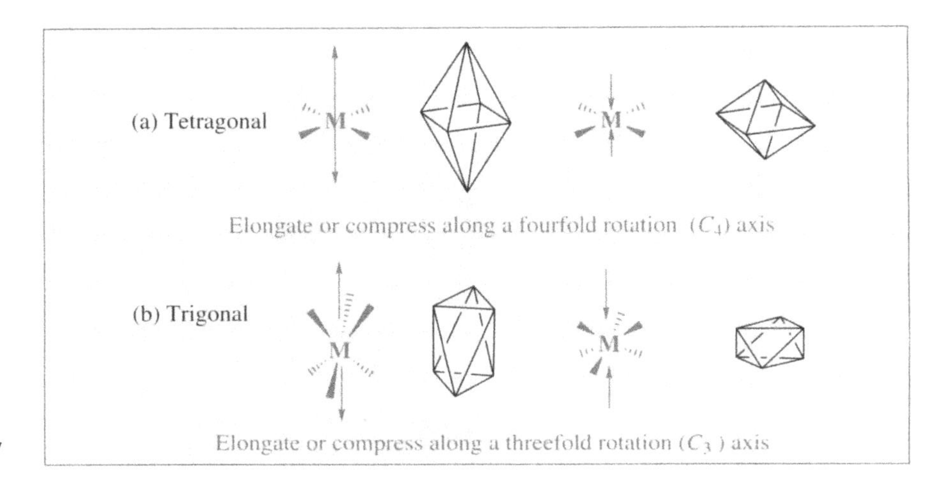

Figure 4.9 Distortions of the octahedral coordination geometry

The higher CNs of 7, 8 and 9 are unusual for d-block metals, although some examples can be found among the complexes of the early second- and third-row metals. The cyano complexes of molybdenum, $[Mo(CN)_8]^{z-}$ (z = 3, 4), provide well-known examples and, in the solid state, can be found with a dodecahedral or a square antiprismatic structure, depending on the counterion present. In contrast, CNs of 8 or 9 are quite typical among the complexes formed by the f-block elements and CNs up to 14 are known. Although regular geometries can be assigned to these higher CNs, the energy differences between the different structures are often small. Since the core-like nature of the f orbitals gives little directional preference in the bonding, the structures of lanthanide complexes tend to be determined by ligand–ligand interactions and distortions from ideal geometries are common. The highest CNs are found with the larger metal ions and the sterically most compact ligands such as NO_2^- or H_2O. An example of 12 coordination is provided by $[Ce(NO_3)_6]^{3-}$, in which six octahedrally disposed nitrate groups each bind through two oxygens to the Ce^{3+} ion to give a distorted icosahedral structure. A rare example of a CN of 14 is found in the solid-state structure of $[U(BH_4)_4]$ in which the U^{4+} ion is bonded to 14 hydrogen atoms.

In addition to the structural isomerism possible for each CN, several other types of isomerism (geometric, optical, linkage, coordination, ligand, ionization, solvate) are possible in transition element complexes.

Geometric Isomerism

In heteroleptic complexes the different possible geometric arrangements of the ligands can lead to isomerism, and important examples of this behaviour are found with CNs 4 and 6. In a square planar complex of formula $[MX_2L_2]$ (X and L are unidentate) the two ligands X may be arranged adjacent to each other, in a *cis* isomer, or opposite each other, in a *trans* isomer. Such isomerism is not possible for a tetrahedral geometry but similar behaviour is possible in CN 6 octahedral complexes of formula $[MX_2L_4]$ in which the two ligands X may be oriented *cis* or *trans* to each other (Figure 4.10a). This type of isomerism also arises when ligands L–L (*e.g.* L–L = $NH_2CH_2CH_2NH_2$) with two donor atoms are present in complexes of formula $[MX_2(L\text{–}L)_2]$. A further example of geometric isomerism can occur in octahedral complexes of formula $[MX_3L_3]$, depending on whether the three ligands X, or L, all lie in the same plane giving a meridional, or *mer*, isomer, or whether they are adjacent forming a triangular face of the octahedron in a facial, or *fac*, isomer (Figure 4.10b).

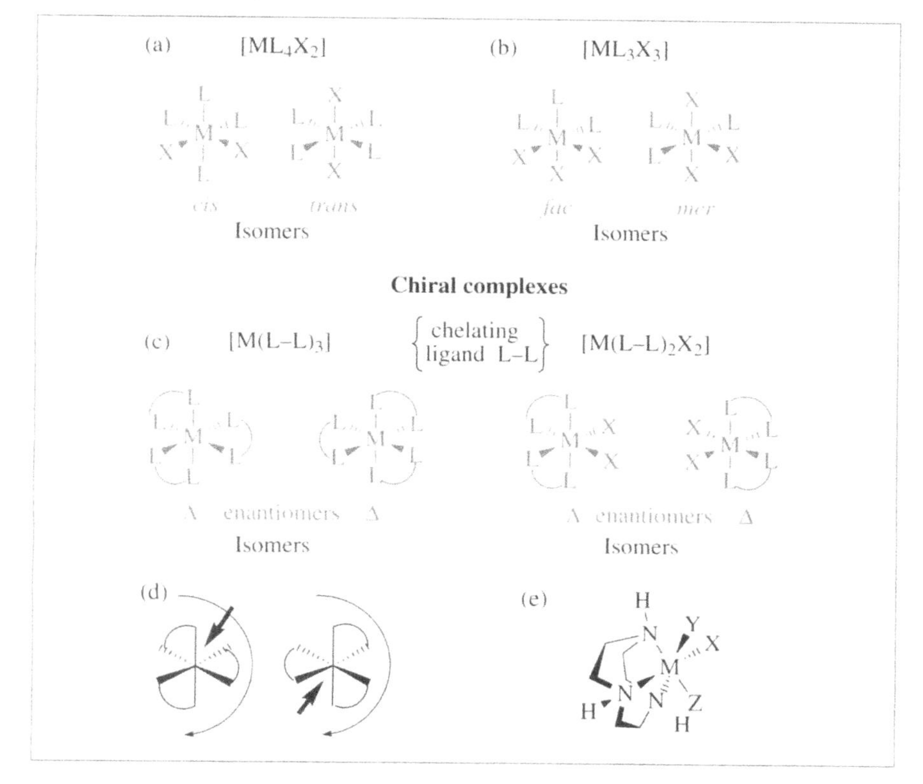

Figure 4.10 Isomers of octahedral complexes

Optical Isomerism

In the octahedral complexes *cis*-$[MX_2(L-L)_2]$ and $[M(L-L)_3]$ the presence of the ligands L–L leads to a further isomeric possibility since the metal centre is chiral (Figure 4.10c). The presence of the L–L ligands gives rise to a propeller shape. When rotated clockwise the Δ isomer appears to screw into the plane while the Λ isomer appears to screw out of the plane (Figure 4.10d). Thus $[Co(en)_3]^{3+}$ is chiral, and can be resolved into its Δ and Λ isomers through fractional crystallization with a chiral counter ion such as tartarate. A tetrahedral complex with four different unidentate ligands will also be chiral, just like a carbon atom in a chiral organic molecule. In practice, examples of this type are rare, but an important example of similar structural type arises with heteroleptic CN 6 complexes containing one *fac*-tridentate ligand and three different unidentate ligands (Figure 4.10e). Some compounds of this type have been found to be effective reagents in the asymmetric synthesis of chiral organic molecules and are of importance in the synthesis of certain fine chemicals and pharmaceuticals.

Linkage Isomerism

Linkage isomerism arises when a ligand may bind to a metal ion through either of two or more different donor atoms. A simple example is provided by complexes containing the thiocyanate ligand, NCS^-, which may bind through the nitrogen to give M–NCS or the sulfur to give M–SCN. The former may be associated with the valence tautomer $^-N{=}C{=}S$ and the latter with $N{\equiv}C{-}S^-$. Jørgensen[1] discovered the first example of such behaviour in the complex $[Co(NH_3)_5(NO_2)]Cl_2$, which can be obtained as a red form, in which the nitrite ligand is bound through oxygen, and as a yellow form, in which the nitrite ligand is bound through nitrogen.

Coordination Isomerism

This form of isomerism arises from the interchange of ligands between cationic and anionic complexes of different metal ions present in a salt. An example is provided by $[Co(NH_3)_6][Cr(CN)_6]$, in which the NH_3 ligands are bound to Co^{3+} and the CN^- ligands to Cr^{3+}, and its coordination isomer $[Cr(NH_3)_6][Co(CN)_6]$ in which the NH_3 ligands are bound to Cr^{3+} and the CN^- ligands to Co^{3+}. A special case of coordination isomerism can arise in which a series of compounds have the same empirical formula but different molecular masses for the salt. This is sometimes referred to as 'polymerization isomerism', although it does not involve polymerization according to a conventional definition involving the linking together of a single repeating unit. An example of polymerization isomerism is provided by the series of salts in which both the cation and the anion contain Co^{3+} and which have the empirical formula $\{Co(NH_3)_3(NO_2)_3\}_x$, *e.g.*

$$[Co(NH_3)_6][Co(NO_2)_6] \ (x = 2)$$

$$[Co(NH_3)_4(NO_2)_2][Co(NH_3)_2(NO_2)_4] \ (x = 2)$$

$$[Co(NH_3)_5(NO_2)][Co(NH_3)_2(NO_2)_4]_2 \ (x = 3)$$

$$[Co(NH_3)_6][Co(NH_3)_2(NO_2)_4]_3 \ (x = 4)$$

Ligand Isomerism

As the name implies, ligand isomerism arises from the presence of ligands which can adopt different isomeric forms. An example is provided by diaminopropane, which may have the amine groups in the terminal 1,3-positions ($H_2NCH_2CH_2CH_2NH_2$) or in the 1,2-positions $\{H_2NCH_2$-$CH(Me)NH_2\}$.

Ionization Isomerism

This arises when the counterion in a complex salt is a proligand and can, in principle, displace a ligand which can then become the counterion. An example is provided by the ionization isomers $[Co(NH_3)_5SO_4]Br$ and $[Co(NH_3)_5Br](SO_4)$.

Solvate Isomerism

This form of isomerism, sometimes known as 'hydrate isomerism' in the special case where water is involved, is similar in some ways to ionization isomerism. Solvate isomers differ by whether or not a solvent molecule is directly bonded to the metal ion or merely present as free solvent in the crystal lattice or solution. An example is provided by the aqua complex $[Cr(H_2O)_6]Cl_3$ and its solvate isomer $[Cr(H_2O)_5Cl]Cl_2.H_2O$.

Worked Problem 4.1

Q Draw all the possible isomers of a five-coordinate complex of formula $[M(L\text{-}L)X_2A]$. In each case, identify the situations in which optical isomers are possible (A and X represent different unidentate ligands; L-L represents a bidentate ligand).

A In five-coordinate complexes the basic regular geometries to consider are square pyramidal and trigonal bipyramidal. The ligands X and A may occupy apical or basal sites in the former and axial or equatorial sites in the latter. Similarly, L–L may span similar locations or different locations in either geometry. The possibilities which arise are shown in Figure 4.11.

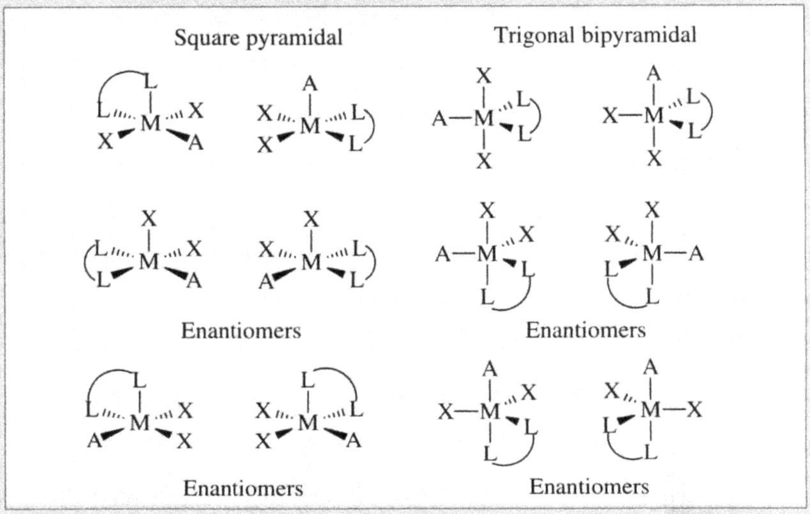

Square pyramidal

Trigonal bipyramidal

Enantiomers

Enantiomers

Enantiomers

Enantiomers

Figure 4.11

4.3 Nomenclature

The systematic naming, or nomenclature, of coordination compounds can be complicated to apply, but it is essential to have some familiarity with the basic rules of nomenclature and to be able to work out the structure of a compound from its systematic name. Only a very brief summary of the rules for naming of coordination compounds can be given here, but more detailed accounts are available elsewhere.[2,3]

4.3.1 Formulae

The formula of a complete coordination entity (*e.g.* $[Co(NH_3)_6]^{3+}$) is written in square brackets. The absence of [—] implies that not all of the ligands in the coordination sphere, typically water or solvent, are given in the formula. As an example, $CoCl_2$ could refer to anhydrous $CoCl_2$, $[CoCl_2(H_2O)_4]$ or $[Co(H_2O)_6]Cl_2$, although the latter might be written as $CoCl_2.4H_2O$ and $CoCl_2.6H_2O$, respectively. The symbol of the central atom is given first followed by the anionic then the neutral ligands, each set in alphabetical order. Where possible the donor atom is written first, one exception being water which is conventionally written H_2O rather than OH_2. Within the formula, brackets should be 'nested' as [{( )}] or [{[( )]}]. Where the formulae of salts are given, cations come before anions, *e.g.* $[Co(H_2O)_6](NO_3)_2$ and $K_3[Fe(CN)_6]$. The oxidation states of the metal ions may be given as superscript Roman numerals following the ion to which they refer, *e.g.* $[Cr^{III}Cl_6]^{3-}$. Abbreviations may be used in formulae. These include standard abbreviations for functional groups, such as Me, Et and Ph respectively for methyl, ethyl or phenyl groups, and specific abbreviations for ligands, *e.g.* py for pyridine and en for ethane-1,2-diamine. Where protic acid ligands are present it is important to remember that the abbreviation for the neutral ligand must include the proton(s), so that ox^{2-} may represent ethanedioate (oxalate) and H_2ox represents ethanedioic acid (oxalic acid).

4.3.2 Names

The names of coordination compounds are constructed by listing the ligands in alphabetical order, regardless of charge, before the name of the central atom. Numerical prefixes are used to indicate the numbers of each ligand present, and the oxidation state is given as Roman numerals in parentheses at the end, *e.g.* dichlorobis(trimethylamine)-platinum(II) for $[PtCl_2(NMe_3)_2]$. Di, tri, *etc.*, are normally used but with more complicated ligand names the prefixes bis, tris, tetrakis, *etc.*, are used to avoid confusion, *e.g.* trichlorotris(trimethylphosphine)rhodium(III) for $[RhCl_3(PMe_3)_3]$. In the names of salts, cations come before anions, as with formulae. Anionic coordination compounds are given the suffix -ate. Either the oxidation state of the central atom {Stock system; *e.g.* hexamminecobalt(III) trichloride for $[Co(NH_3)_6]Cl_3$}[4] or the charge on the complex (Ewens–Bassett system)[5] may be given, *e.g.* tripotassium hexacyanoferrate(3–) for $K_3[Fe(CN)_6]$; in the Stock system this is tripotassium hexacyanoferrate(III). Note that ammine in this context refers to NH_3 as a ligand, the term -amine being used for a derivative such as dimethylamine, $NHMe_2$. Stereochemical descriptors may precede names or formulae to indicate which isomeric form is present, *e.g.* *cis*-diamminedichloroplatinum(II) for *cis*-$[PtCl_2(NH_3)_2]$.

4.3.3 Special Symbols

Greek letters are used as special symbols to signify particular structural features in a compound. The character μ is used to denote an atom or group which bridges between two or more metal centres. A subscript denotes the number of atoms bridged as shown in **4.1** and **4.2**. The symbol η is used to denote connected atoms in a ligand which bind to the metal atom and a superscript identifies the number involved, as shown

$[(ML_2)(\mu_2\text{-}Cl)_2]$

4.1

$[(ML_2)(\mu_2\text{-}Cl)_3(\mu_3\text{-}O)]$

4.2

$[PdCl_2(\eta^2\text{-}C_2H_4)]$

4.3

$[Fe(\eta^5\text{-}C_5H_5)I(CO)_2]$

4.4

$[CoCl_3\{(NH_2CH_2CH_2)_2NH\text{-}\kappa^3\text{-}N,N,N\}]$

4.5

for **4.3** and **4.4.** Finally, the symbol κ is used to denote unconnected atoms in a polydentate ligand which bind to the metal centre and again a superscript denotes the number of donor atoms involved and may be followed by the appropriate donor atom symbol italicized, as in **4.5**.

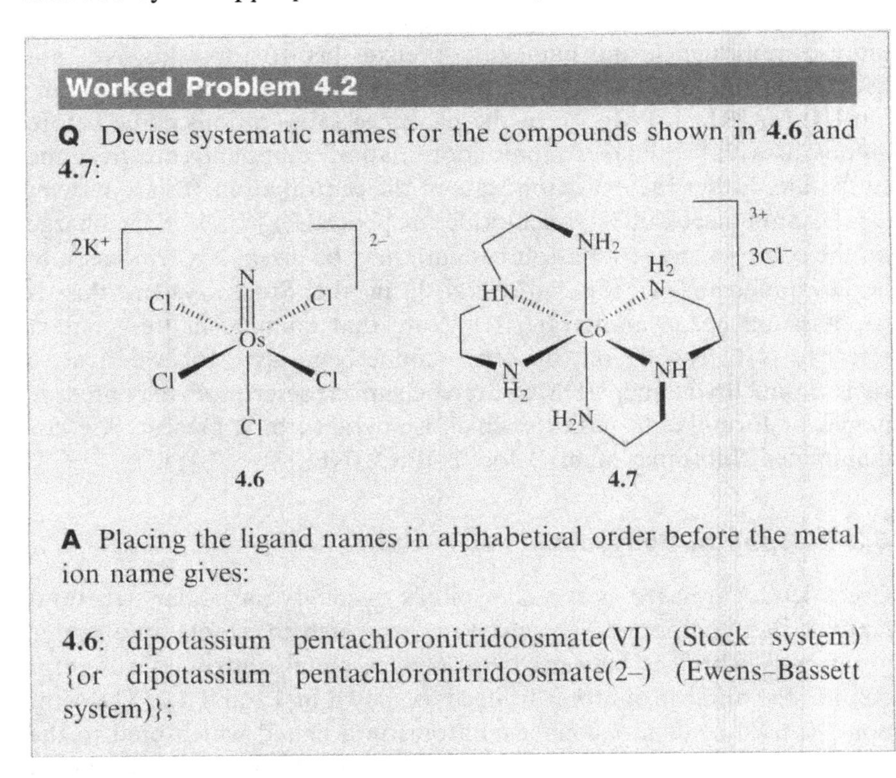

Worked Problem 4.2

Q Devise systematic names for the compounds shown in **4.6** and **4.7**:

4.6 **4.7**

A Placing the ligand names in alphabetical order before the metal ion name gives:

4.6: dipotassium pentachloronitridoosmate(VI) (Stock system) {or dipotassium pentachloronitridoosmate(2–) (Ewens–Bassett system)};

4.7: bis(1,4,7-triazaheptane-κ^3-N,N',N'')cobalt(III) trichloride.

Summary of Key Points

1. *The environment of a metal ion in a coordination compound* is defined by its coordination number and coordination geometry.

2. *Any of a wide variety of ligands* may be involved in forming coordination compounds with metals.

3. *Different isomeric structures* are possible for many complexes and seven different types of isomerism may be identified.

4. *A systematic system of nomenclature* exists for coordination compounds of metal ions.

Problems

4.1. Consult text books from the Further Reading section and, for each of the coordination geometries for coordination numbers from 2 to 8, identify at least one known compound which exemplifies that coordination geometry. Draw the structure of each using a notation which reveals the three-dimensional arrangement of the groups.

4.2. List the seven types of isomerism possible for coordination compounds, giving an example of each.

4.3. Draw all the possible isomers of:

(i) a four-coordinate complex $[ML_2XY]$
(ii) an octahedral complex $[MA_2LXYZ]$
In each case, identify the situations in which optical isomers are possible (A, L, X, Y, Z represent different unidentate ligands)

4.4. Draw the possible isomers of the complex $[Ru\{S_2C_2(CF_3)_2\}(CO)(PPh_3)_2]$, in which $S_2C_2(CF_3)_2$ is a bidentate dithiolate ligand (Figure 4.1).

4.5. Devise systematic names for the compounds shown in Formulae **4.8**–**4.11**:

4.8

4.9

4.10

4.11

References

1. S. M. Jørgensen, *Z. Anorg. Chem.*, 1893, **5**, 147; 1899, **19**, 109.
2. J. E. Huheey, E. A. Keiter and R. L. Keiter, *Inorganic Chemistry*, 4th edn., HarperCollins, New York, 1993, appendix A3.
3. A comprehensive account may be found in the chapter by T. E. Sloane in *Comprehensive Coordination Chemistry*, ed. G. Wilkinson, R.D. Gillard and J.A. McCleverty, Pergamon Press, Oxford, 1987.
4. A. Stock, *Z. Angew. Chem.*, 1919, **27**, 373.
5. R. G. V. Ewens and H. Bassett, *Chem. Ind.*, 1949, **27**, 131.

5
The Thermodynamics of Complex Formation

Aims

By the end of this chapter you should understand the terms:

- Hydration enthalpy
- Stability or formation constant
- Overall and stepwise stability constants
- Chelate effect
- Macrocyclic effect
- Preorganization
- Equilibrium template effect
- Kinetic template effect
- Self-assembly

and have an understanding of:

- The rôles of enthalpy and entropy in the thermodynamics of complex formation
- The relationship between stability constants and electrode potentials for complexes in different oxidation states
- The terms *hard* and *soft* as applied to metal ions

and a knowledge of some examples of:

- Metal templated reactions
- Metal-directed self-assembly reactions

The chapter will assume a knowledge of basic thermodynamics, including the meaning of the terms equilibrium constant, enthalpy, entropy, free energy and electrode potential, and the relationships between them.

5.1 Introduction

One of the first questions one might ask about forming a metal complex is: how strong is the metal ion to ligand binding? In other words, what is the equilibrium constant for complex formation? A consideration of thermodynamics allows us to quantify this aspect of complex formation and relate it to the electrode potential at which the complex reduces or oxidizes. This will not be the same as the electrode potential of the simple solvated metal ion and will depend on the relative values of the equilibrium constants for forming the oxidized and reduced forms of the complex. The basic thermodynamic equations which are needed here show the relationships between the standard free energy ($\Delta G^{\ominus}$) of the reaction and the equilibrium constant (K), the heat of reaction, or standard enthalpy ($\Delta H^{\ominus}$), the standard entropy ($\Delta S^{\ominus}$) and the standard electrode potential ($E^{\ominus}$) for standard reduction of the complex (equations 5.1–5.3).

$$\Delta G^{\ominus} = -RT\ln K \tag{5.1}$$

$$\Delta G^{\ominus} = \Delta H^{\ominus} - T\Delta S^{\ominus} \tag{5.2}$$

$$\Delta G^{\ominus} = -nFE^{\ominus} \tag{5.3}$$

where R is the gas constant, F the Faraday constant, n the number of electrons involved in the reduction of the complex and T the temperature in Kelvin.

At a deeper level, studies of the thermodynamics of complex formation reveal insights into metal–ligand bonding and contribute to an understanding of the ways in which ligand structure relates to the strength of ligand binding, and hence to the electrode potential of the complex. Such knowledge allows complexes with particular properties to be designed to suit particular applications. Throughout evolution, biological systems have exploited the particular properties of metal ions and so have developed molecules which bind metal ions very strongly and, in some cases, can control the redox potentials of the resulting metal complex. Another important feature of transition element chemistry in solution is the ability of the metal ion to bring together several ligands in a single complex. This has implications for synthesis in that metal ions may be used to assemble polynuclear structures, control the structures of reacting ligands and possibly activate ligands towards reaction.

5.2 The Thermodynamics of Complex Formation

5.2.1 Hydration Enthalpies

Syntheses of transition element complexes are normally carried out in solution, often in water, and so involve the reactions of solvated metal ions. The association of the solvent molecules with an unsolvated metal ion releases energy and, for aqueous solutions, a **hydration enthalpy** is associated with this process. The hydration enthalpies of first-row d-block metal dications (Figure 5.1) increase in magnitude with increasing atomic number and decreasing ionic radius. The trend is not completely uniform for reasons which will be explained in Chapter 6, but the overall appearance of the plot is similar to that of the lattice enthalpies of metal dichlorides against atomic number shown in Figure 3.3. The hydration enthalpies of the lanthanide trications also increase with increasing atomic number, but pursue a more regular curve in following their decreasing ionic radii (Figure 5.2). This increase in hydration enthalpy with decreasing ionic radius reflects stronger binding of the solvent molecules to the smaller more polarizing cations.

The **hydration enthalpy**, $\Delta H^{\ominus}_{hyd}$, of a metal ion can be defined as the standard enthalpy change when one mole of unsolvated free metal ions in the gaseous state is converted to one mole of hydrated metal ions in aqueous solution. Since this will be an exothermic process, $\Delta H^{\ominus}_{hyd}$ will have a negative value. The **hydration energy** is the standard molar energy change for this process.

The dissolution of an anhydrous metal salt in a solvent such as water involves the disruption of the crystal lattice of the salt, a process which will consume energy depending on the lattice energy of the salt. Assuming that the salt is fully dissociated in the solution, dissolution in water produces fully hydrated metal ions and releases hydration energy through forming the aquated metal ion. There is also a contribution from the solvation of the counterions. If the energy available from the solvation of the anion and cation exceeds the lattice energy, a salt can dissolve in

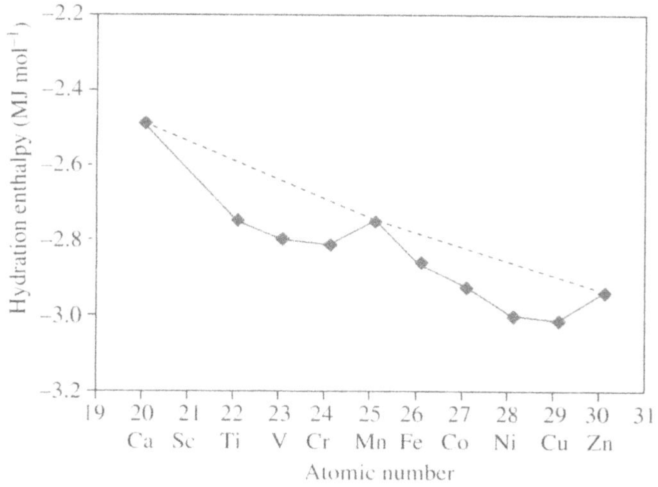

Figure 5.1 The variation in hydration enthalpy of M^{2+} ions across the first row of the d-block

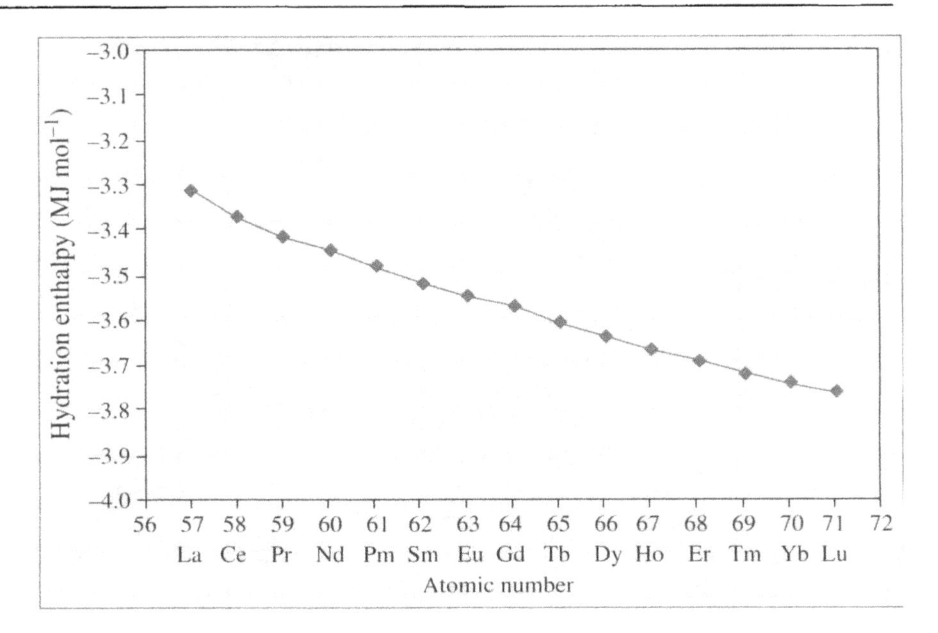

Figure 5.2 The variation in hydration enthalpy of Ln^{3+} ions across the lanthanide series

water to form a solution of hydrated metal ions. Often the enthalpy of dissolution is insufficient to overcome the lattice enthalpy, but the favourable entropic consequences of converting the salt from the solid to the solution phase can allow dissolution to proceed. In such cases the heat deficit is made up from the internal energy of the system, and the solution cools as the salt dissolves.

In a chemical reaction, the relative concentrations of reactants and products which co-exist at equilibrium may be defined by an **equilibrium constant**, K. As an example, in a reaction where two reactants A and B combine to form two products C and D, the value of K is given by:

$$K = \frac{\gamma_C \gamma_D [C][D]}{\gamma_A \gamma_B [A][B]}$$

Here, square brackets denote the **concentrations** of A, B, C and D and γ_A, γ_B, γ_C and γ_D denote their respective *activity coefficients*. These coefficients are present to take account of the non-ideal behaviour of real systems, but, in practice, the values of γ_A, γ_B, γ_C and γ_D are difficult to measure and are not usually known. In dilute solutions of high ionic strength they are often assumed to be 1 so that, to a first approximation, the value of K is defined by the concentrations of A, B, C and D alone.

5.2.2 Equilibrium Constants for Complex Formation

As with any other chemical reaction, the formation of a metal complex from a metal ion and a set of proligands can be described by an equilibrium constant. In its simplest form, a complexation reaction might involve the reaction of unsolvated metal ions in the gas phase with gas phase proligands to form a complex. In practice it is difficult to study such reactions in the gas phase and complex formation is normally studied in solution, often in water. This introduces the complication that the solvent can also function as a ligand, so that complex formation will involve the displacement of solvent from the metal coordination sphere by the proligand.

In complexation reactions the solvent is normally present in large excess so that its concentration is essentially constant. Thus, in cases where the reactivity of the coordinated solvent is not an issue, it is possible to simplify considerations of complex formation by omitting the solvent from the description of the equilibrium. In such cases the formation of a metal complex from a metal ion and one or more proligands, L, according to equation 5.4 can be described by the simplified equation 5.5, in which the solvent and the charge on the complex are omitted:

$$[M(H_2O)_x]^{z+}{}_{(aq)} + L_{(aq)} \rightleftharpoons [M(H_2O)_{(x-1)}(L)]^{z+}{}_{(aq)} + H_2O_{(aq)} \quad (5.4)$$

$$M + L \rightleftharpoons ML \quad (5.5)$$

The equilibrium constant for the process shown in equation 5.5 is called a 'stability constant'. The terms 'formation constant' and 'binding constant' are also sometimes used in this context. The stability constant can be defined on the basis of concentrations assuming unit activity coefficients, an assumption which is not usually too unrealistic in dilute solutions of high ionic strength. The stability constant, K_1, for the reaction shown in equation 5.5 is given by equation 5.6:

Unfortunately, **square brackets** have conventionally been used both to denote concentrations, [A] representing the concentration of A in mol dm^{-3}, and coordination complexes, *e.g.* $[Co(NH_3)_6]^{3+}$. This presents a potential source of confusion when defining equilibrium constants for coordination compounds since often only one set of brackets is used. Usually the function of the square bracket is clear from the context in which it is used, but for clarity in this text, two sets of square brackets will be used when appropriate. Since equilibrium constants are dimensionsless it is the convention to divide concentrations in an equilibrium constant expression by a standard concentration, mol dm^{-3}.

$$K_1 = [ML]/[M][L] \quad (5.6)$$

Similarly, for subsequent additions of L to M, further stability constants K_2 to K_n may be defined according to equations 5.7 and 5.8:

for $ML + L \rightleftharpoons ML_2$, K_2 is defined by:

$$K_2 = [ML_2]/[ML][L] \quad (5.7)$$

and so on to $ML_{(n-1)} + L \rightleftharpoons ML_n$ with:

$$K_n = [ML_n]/[ML_{(n-1)}][L] \quad (5.8)$$

The equilibrium constants K_1 to K_n are called stepwise stability constants and represent the equilibria involved in the stepwise addition of one ligand to the metal. However, equilibrium constants could also be written for an overall reaction involving more than one ligand. These are known as overall stability constants and are typically signified by the symbol β with a suffix to indicate the number of ligands involved, as shown in equations 5.9–5.11:

$$\text{for } M + L \rightleftharpoons ML \qquad \beta_1 = [ML]/[M][L] \quad (5.9)$$

$$\text{for } M + 2L \rightleftharpoons ML_2 \qquad \beta_2 = [ML_2]/[M][L]^2 \quad (5.10)$$

and so on to

$$M + nL \rightleftharpoons ML_n \qquad \beta_n = [ML_n]/[M][L]^n \quad (5.11)$$

The value of an overall stability constant β_j is the product of the stepwise stability constants K_1 to K_j as shown by equation 5.12:

$$\beta_j = K_1K_2 \ldots K_j \quad \text{or} \quad \beta_j = \prod_{i=1}^{i=j} K_j \tag{5.12}$$

This relationship between β_n or K_n values means that either may be used in equation 5.1 but β_n must be used for reactions of the type defined by equations 5.9–5.11 and K_n for reactions of the type defined by equations 5.6–5.8.

Normally, the magnitudes of K_n decrease with increasing n. In part this is the result of statistical factors, but the addition of ligands may also influence the electronic properties of the metal and so the binding of subsequent ligands. Steric interactions between ligands, or decreasing positive charge on the complex if the ligands are negatively charged, can also influence the stability constant. Where the stepwise addition of ligands occurs, several metal complexes may co-exist in solution. The concentration of each will depend upon the relative proportions of metal and ligand present, an example being shown in Figure 5.3. The measurement of stability constants is often carried out using spectrophotometric or electrochemical measurements (Box 5.1). Such measurements are more difficult when more than one species is present since the spectroscopic features of one complex may be obscured by those of another.

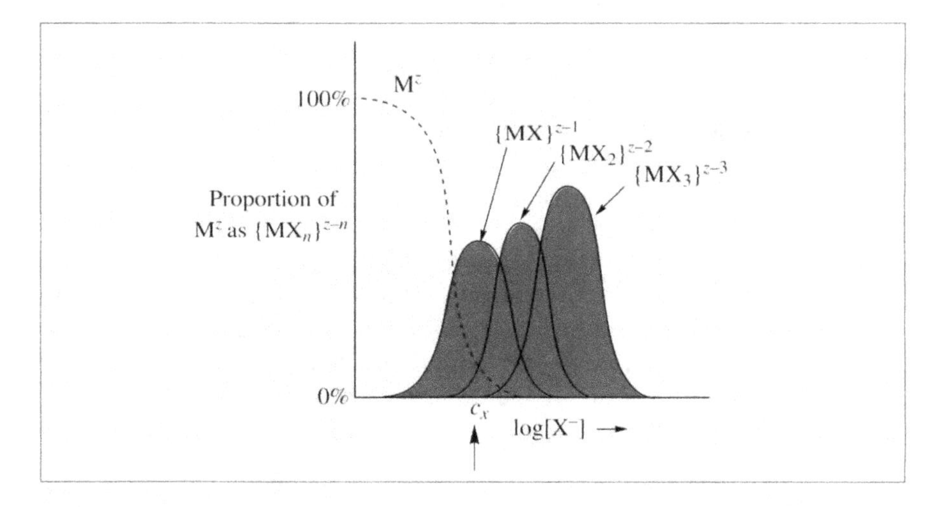

Figure 5.3 A schematic diagram showing how the distribution of a metal ion, M^{z+}, among different complexes, $\{MX_n\}^{(z-n)}$, varies with the concentration of X^-, $[X^-]$. At a particular value of $[X^-]$, more than one complex may be present in solution at equilibrium

Box 5.1 Measuring the Stability Constant of a Metal Complex

There are many methods for determining stability constants but measurements of electrode potentials, pH titrations or the use of spectroscopic methods for determining the concentrations of species in solution are of particular importance. In the simplest case, where only one complex is formed as represented by equation 5.6, if the initial concentrations of added M and L are known it is only necessary to determine the concentration of ML or free M or free L in the solution at equilibrium to determine K. If more than one complex is formed, the situation becomes more complicated. To determine n formation constants in the M + L system, $n + 2$ independent concentration measurements are needed. Provided that the initial concentrations of M and L are known, n further measurements will be required. Concentration measurements of M or L as a function of varying [M] and [L] ratios, followed by a computer-aided analysis, provides a means of measuring β_n values.

Worked Problem 5.1

Q At 298 K a 0.05 mol dm^{-3} solution of metal ions, M^{z+}, in water has an absorbance of 0.6 at the wavelength, λ_{max}, of maximum optical absorption for aquated M^{z+}. A similar solution initially containing 0.1 mol dm^{-3} of M^{z+} and 0.08 mol dm^{-3} of a polydentate ligand L has an absorbance of 0.3 at wavelength λ_{max} and 298 K. Assuming that $\{ML\}^{z+}$ does not absorb light at wavelength λ_{max}, and that $\{ML\}^{z+}$ is the only complex formed from M^{z+} and L, calculate the equilibrium constant for formation of $\{ML\}^{z+}$ from M^{z+} and L in aqueous solution at 298 K.

A Assume that the Beer–Lambert law applies. This states that $A = \varepsilon cl$, where A = absorbance, ε = molar absorptivity (dm^3 mol^{-1} cm^{-1}), l = path length (cm) and c = concentration (mol dm^{-3}). In this example, ε and l are constant.

If the absorbance of a 0.05 mol dm^{-3} solution of M^{z+} is 0.6, then $0.6 = 0.05\varepsilon l$, so that $\varepsilon l = 0.6/0.05 = 12$. If the absorbance of the mixture of M^{z+} and L is 0.3 for the same λ_{max} and l values and no other species absorb at λ_{max}, the free M^{z+} concentration in the mixture is given by $0.3 = [M^{z+}]\varepsilon l = 12[M^{z+}]$, so that $[M^{z+}] = 0.3/12 = 0.025$ mol dm^{-3}. The M^{z+} which is not present as free M^{z+} must be in the form $\{ML\}^{z+}$, as this is the only complex formed. Hence the concentration of $\{ML\}^{z+}$ is $0.1 - 0.025 = 0.075$ mol dm^{-3}. Since we started with an L concentration of 0.08 mol dm^{-3}, of which 0.075 mol dm^{-3} is consumed in forming ML, the remaining concentration of free L is $0.08 - 0.075 = 0.005$ mol dm^{-3}. We now have all the concentrations needed to calculate the stability constant, K_1, for forming $\{ML\}^{z+}$ under the conditions used:

$$K_1 = [\{ML\}^{z+}]/[M^{z+}][L] = 0.075/(0.025 \times 0.005) = 600$$

The terms 'stable' and 'unstable' are often used in a general way to describe the reactivity of compounds. In the case of metal complexes these terms have a specific meaning which relates to the thermodynamics of complex formation. *Stable* complexes might be defined as those which have overall stability constants greater than one, so a negative free energy for formation, under the particular conditions being considered. *Unstable* complexes would then have overall stability constants less than one, so a positive free energy for formation. Thus solutions of stable complexes would not be expected to react with the solvent, or any other proligands present, to form thermodynamically more stable complexes. In contrast, unstable complexes might be expected to undergo ligand substitution reactions to form a more thermodynamically stable species.

5.2.3 Hard and Soft Acids and Bases

A large number of stability constants have been measured and certain general trends emerge from the results obtained. One is that particular types of ligand donor atom form stronger complexes with certain metal ions. That is, their complexes with certain metal ions have higher stability constants than their complexes with other metal ions. This allows a broad classification of metal ions according to the type of ligands with which they form the strongest complexes. Those metals which form their most stable complexes with oxygen donor atoms are called class a, or hard, metal ions and those which form their strongest complexes with sulfur or phosphorus donor are called class b, or soft, metal ions. There is also a group of metal ions which show borderline behaviour. Class a metal ions are usually smaller, more highly charged, cations. They form their strongest complexes with hard bases containing the smaller electronegative donor atoms O, N or F. At the other end of the scale are the soft metal ions, *i.e.* the larger more polarizable metal ions, often in their lower oxidation states. These are soft acids and form their strongest complexes with soft bases which contain larger more polarizable and less electronegative donor atoms such as S, Se, P or As. This classification is summarized in Table 5.1. The hard/soft classification of metal ions and ligand donor atoms is a useful qualitative concept in that it is a guide to predicting which ligands may be more suited to forming complexes with a particular metal ion. Typically phosphine, R_3P (R = hydrocarbyl), or thiolate, RS^-, ligands stabilize lower oxidation states such as Rh^+, Ir^+ or Cu^+, whereas fluoride or oxide ions are better suited to higher oxidation state metal ions, allowing, for example, the isolation of the Co^{4+} complex $[CoF_6]^{2-}$. In some applications of metal sequestering agents there is a need for ligands which show very large stability constants for particular metals. Examples are provided by the use of pyridinones such as **5.1** as chelating agents to remove iron from patients suffering iron overload as a result of repeated blood transfusions in the treatment of β-thalassaemia. Either D-penicillamine (**5.2**) or the tetramine proligand **5.3** can be used to remove copper from patients suffering from Wilson's disease, a hereditary complaint which leads to a build-up of copper in the liver. Oxime reagents are also used to bind copper in its purification by solvent extraction using so called LIX reagents, *e.g.* **5.4**.

Table 5.1 Examples of class a (hard) and class b (soft) metal ions and ligands

Class a (hard)	*Borderline*	*Class b (soft)*
Metal ions	**Metal ions**	**Metal ions**
Mn^{2+}, Sc^{3+}, Cr^{3+}, Fe^{3+}, Ti^{4+}, Sc^{3+}	Fe^{2+}, Co^{2+}, Ni^{2+}, Cu^{2+}	Hg^+, Hg^{2+}, Cu^+, Ag^+, Au^+, Pd^{2+}, Pt^{2+}, Rh^+, Ir^+
Ligands	**Ligands**	**Ligands**
F^-, R_2O, ROH, OH^-, RCO_2^-, SO_4^{2-}, NR_3, Cl^-, $SC\mathit{N}^-$ [a]	Br^-, py	R_2S, R_3P, R_3As, RNC, CO, CN^-, I^-, NCS^- [a]

[a]In the case of ambidentate ligands, the donor atom is italicized

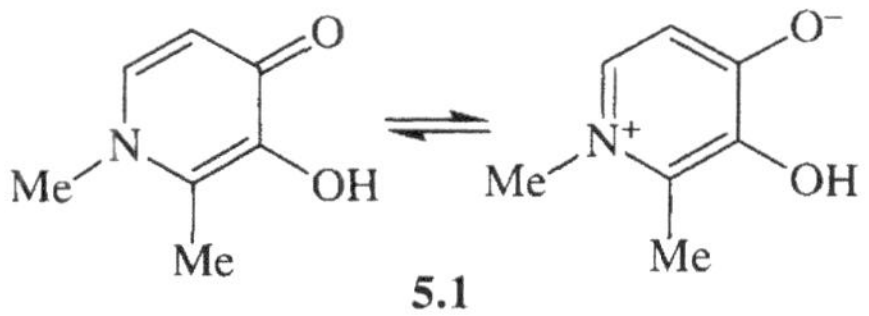

5.1 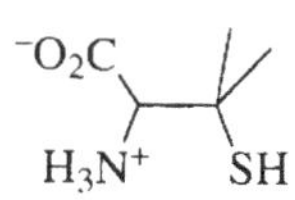 **5.2** 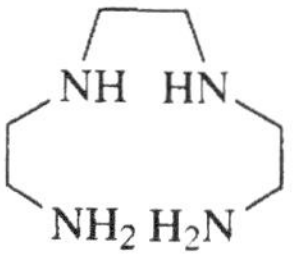 **5.3**

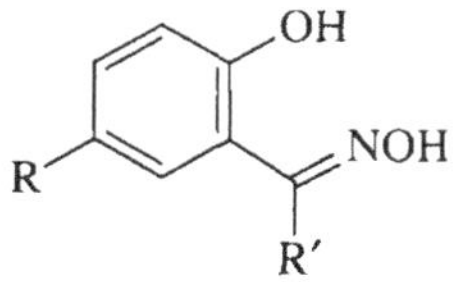

R = alkyl, R′ = H, alkyl, aryl
5.4

5.2.4 The Contributions of Enthalpy and Entropy

When a solvated metal ion reacts with a proligand, which could be polydentate, to form a complex, several things must happen. Firstly, coordinated solvent must be removed from the metal ion coordination sphere to create a space for the incoming ligand, and the solvation shell around the metal must be reorganized to accommodate the new structure. This will require an energy input $\Delta G^{\ominus}_{SM}$. A similar process will occur around the proligand, requiring some energy input $\Delta G^{\ominus}_{SL}$. The energies required to effect these desolvation processes will contribute to the thermodynamics of the complexation reaction. Two further processes also need to be considered. A polyatomic ligand may need to rearrange its structure in order to adopt a conformation suitable for binding to the metal, so absorbing some conformational energy $\Delta G^\ominus_{CL}$ Finally, the formation of the metal to ligand donor atom bonds should release energy and contribute $\Delta G^\ominus_{ML}$ to the driving force for the reaction. Thus the free energy of the reaction may be described as the sum of these various components according to equation 5.13:

$$\Delta G^\ominus = \Delta G^\ominus_{SM} + \Delta G^\ominus_{SL} + \Delta G^\ominus_{CL} + \Delta G^\ominus_{ML} \tag{5.13}$$

Each of the component free energy terms of equation 5.13 may be divided into enthalpic contributions ($\Delta H^\ominus$) and entropic contributions ($T\Delta S^\ominus$) according to equation 5.2. Usually it might be anticipated that the desolvation energies, $\Delta G^\ominus_{SM}$ and $\Delta G^\ominus_{SL}$, would contain positive enthalpy contributions, since bonding interactions between the metal ion and solvent molecules must be disrupted, together with positive entropy terms (hence a negative $-T\Delta S^\ominus$), as the disorder of the system increases on separating solvent molecules from the metal ion or proligand. In the case of $\Delta G^\ominus_{CL}$ it might be expected that there would be a positive contribution from enthalpy, since bonds may need to be rotated into a sterically less favourable structure, and the entropy term would be negative, as the structure of the proligand becomes more ordered. In the case of $\Delta G^\ominus_{ML}$, a negative enthalpy contribution should arise from the formation of the metal–ligand bond, but again a negative entropy contribution would result from the increasing order associated with combining metal ion and proligand. The balance between these various enthalpic and entropic contributions will vary from complex to complex and can have important consequences for complex formation.

5.2.5 The Chelate Effect

The stability constants of complexes containing chelating ligands are found to be higher than for their non-chelated counterparts. This observation is known as the chelate effect, and an example is provided by the value of $10^{10.6}$ for β_2 of $[Cd(en)_2]^{2+}$ formation, which is some 10^4 times larger than that of $10^{6.52}$ for β_4 of $[Cd(NH_2Me)_4]^{2+}$ formation. At a qualitative level the chelate effect may be explained by observing that, with a unidentate ligand, dissociation leads to the complete loss of the ligand. In contrast, when one end of a chelated ligand dissociates, the ligand remains attached to the metal ion through the other donor group so that it is more likely that the dissociated end will re-attach than that the chelate ligand is displaced by a second ligand. A consideration of the thermodynamics of forming these complexes reveals the entropy term as the driving force for the chelate effect (Box 5.2). In general it appears that, in relatively simple cases such as this, entropy is the most probable cause of the chelate effect. In the example of the cadmium complexes, four Cd–N bonds are present in each case, and the main difference is the entropic effect of releasing two additional molecules when replacing four $MeNH_2$ by two en ligands.

Box 5.2 The Chelate Effect

An example of the chelate effect is provided by the reaction between cadmium(2+) ions and the unidentate proligand methylamine (equation 5.14) or the chelating bidentate proligand ethane-1,2-diamine, en (equation 5.15). The stability constant for the formation of $[Cd(MeNH_2)_4]^{2+}$ in aqueous media is $10^{6.52}$, but for the formation of the chelate analogue $[Cd(en)_2]^{2+}$ the stability constant is $10^{10.6}$, some four orders of magnitude larger. The origin of this difference can be seen in the enthalpy and entropy values for the two reactions. Both reactions have similar $\Delta H^\ominus$ values, so the large difference in free energy comes mainly from the $T\Delta S^\ominus$ contribution. Another way to analyse these data is to consider the direct replacement of four methylamine ligands by two en. The equation for this reaction can be generated by subtracting equation 5.14 from equation 5.15 to give equation 5.16. The values of $\log_{10}\beta_n$, $\Delta G^\ominus$, $\Delta H^\ominus$ and $\Delta S^\ominus$ may be manipulated following the same logic to provide values for equation 5.16:

$$Cd^{2+}_{(aq)} + 4MeNH_{2(aq)} \rightleftharpoons [Cd(MeNH_2)_4]^{2+}_{(aq)} \qquad (5.14)$$

$\Delta G^\ominus = -37.2\ kJ\ mol^{-1}$ and $\log_{10}\beta_n = 6.52$
$\Delta H^\ominus = -57.3\ kJ\ mol^{-1}$ and $\Delta S^\ominus = -67.3\ J\ K^{-1}\ mol^{-1}$
$-T\Delta S^\ominus = 20.1\ kJ\ mol^{-1}$

$$Cd^{2+}_{(aq)} + 2en_{(aq)} \rightleftharpoons [Cd(en)_2]^{2+}_{(aq)} \qquad (5.15)$$

$\Delta G^\ominus = -60.7\ kJ\ mol^{-1}$ and $\log_{10}\beta_n = 10.6$
$\Delta H^\ominus = -56.5\ kJ\ mol^{-1}$ and $\Delta S^\ominus = +14.1\ J\ K^{-1}\ mol^{-1}$
$-T\Delta S^\ominus = -4.2\ kJ\ mol^{-1}$
(5.15) – (5.14) = (5.16)

$$[Cd(MeNH_2)_4]^{2+}_{(aq)} + 2en_{(aq)} \rightleftharpoons [Cd(en)_2]^{2+}_{(aq)} + 4MeNH_{2(aq)} \qquad (5.16)$$

Values for equation 5.16 are:

$\log_{10}K = 10.6 - 6.52 = 4.08$
$\Delta G^{\circ} = -60.7 - (-37.2) = -23.5$ kJ mol^{-1}
$\Delta H^{\circ} = -56.5 - (-57.3) = +0.8$ kJ mol^{-1}
$\Delta S^{\circ} = +14.1 - (-67.3) = 81.4$ J K^{-1} mol^{-1}
$-T\Delta S^{\circ} = -4.2 - 20.1 = -24.3$ kJ mol^{-1}

These figures show that the replacement of the non-chelated ligand is slightly opposed by the enthalpy contribution and entirely driven by the $-T\Delta S^{\circ}$ contribution.

Worked Problem 5.2

Q Given that, for the reaction in equation 5.17, $\Delta H^{\circ} = -46.47$ kJ mol^{-1} and $\Delta S^{\circ} = -8.37$ J K^{-1} mol^{-1} and, for the reaction in equation 5.18, $\Delta H^{\circ} = -54.43$ kJ mol^{-1} and $\log_{10}K = 10.72$, calculate $\log_{10}K$ at 298 K for the reaction in equation 5.19 and comment on the driving force for this reaction (en = $H_2NCH_2CH_2NH_2$).

$$[Cu(OH_2)_4]^{2+}{}_{(aq)} + 2NH_{3(aq)} \rightleftharpoons [Cu(NH_3)_2(OH_2)_2]^{2+}{}_{(aq)} + 2H_2O_{(aq)} \quad (5.17)$$

$$[Cu(OH_2)_4]^{2+}{}_{(aq)} + en_{(aq)} \rightleftharpoons [Cu(en)(OH_2)_2]^{2+}{}_{(aq)} + 2H_2O_{(aq)} \quad (5.18)$$

$$[Cu(NH_3)_2(OH_2)_2]^{2+}{}_{(aq)} + en_{(aq)} \rightleftharpoons [Cu(en)(OH_2)_2]^{2+}{}_{(aq)} + 2NH_{3(aq)} \quad (5.19)$$

A In order to calculate $\log_{10}K$ for the reaction in equation 5.19 it is necessary to first obtain ΔG°. To comment on the driving force it will also be necessary to know the relative contributions of the enthalpy (ΔH°) and entropy ($-T\Delta S^{\circ}$) terms to ΔG° and hence K. These values can be calculated using the data provided and equations 5.1 and 5.2.

For equation 5.17, $\Delta H^{\circ} = -46.47$ kJ mol^{-1} and $\Delta S^{\circ} = -8.37$ J K^{-1} mol^{-1}; so, at 298 K, $\Delta G^{\circ} = -46.47 - (-8.37 \times 298/1000) = -43.98$ kJ mol^{-1} and $\log_{10}K = 7.707$.

Similarly for equation 5.18, $\Delta H^{\circ} = -54.43$ kJ mol^{-1} and $\log_{10}K = 10.72$, so, at 298 K, $\Delta G^{\circ} = -61.17$ kJ mol^{-1} and $\Delta S^{\circ} = +22.6$ J K^{-1} mol^{-1}.

Thus values for equation 5.19 are obtained by subtracting those for equation 5.17 from those for equation 5.18 as follows:

$\Delta H^\ominus = -54.43 - (-46.47) = -7.96$ kJ mol^{-1}
$\Delta S^\ominus = +22.6 - (-8.37) = +30.97$ J K^{-1} mol^{-1}
$\Delta G^\ominus = -61.16 - (-43.98) = -17.18$ kJ mol^{-1}
$\log_{10}K = 10.72 - 7.707 = 3.013$

These values show a contribution to $\Delta G^\ominus$ of -7.96 kJ mol^{-1} from $\Delta H^\ominus$ and of -9.23 kJ mol^{-1} from the $-T\Delta S^\ominus$ term. Thus the entropy and enthalpy terms make a similar contribution to the driving force. The enthalpy may be in part associated with the more basic, better donor, character of the alkylamine group in en compared to NH_3. The entropy benefit may be related to the replacement of two NH_3 ligands by one en ligand.

5.2.6 The Macrocyclic Effect

The finding that the stability constants of complexes of macrocyclic ligands tend to be higher than those of their acyclic counterparts is known as the macrocyclic effect. As an example, the Ni^{2+} complex of the macrocyclic ligand cyclam has a stability constant some 10^7 times larger than that of its counterpart formed with the acyclic ligand '2,3,2-tet' (Figure 5.4a). It might be thought that entropy would provide a driving force for the macrocyclic effect, as it does for the chelate effect, but the situation is actually less clear-cut for the macrocyclic effect. A consideration of the thermodynamics of complex formation reveals that the enthalpy of forming $[Ni(cyclam)]^{2+}$ is substantially more negative than that of forming $[Ni(2,3,2\text{-tet})]^{2+}$, and the entropic contribution to $\Delta G^\ominus$ ($-T\Delta S^\ominus$) is actually smaller for the macrocyclic ligand. However, in the case of the Cu^{2+} complexes of cyclen and its acyclic counterpart 2,2,2-tet, it is the entropy component which is the major contributor to the macrocyclic effect. The enthalpy of forming the macrocyclic complex is smaller than for the acyclic ligand. Two major factors are at work here. The first is the match between the size of the macrocycle cavity and the metal ion diameter. If these are closely matched, metal–ligand bonding is optimized and a favourable enthalpy of formation results, as with Ni^{2+} and cyclam. If the cavity is too small for the metal ion to fit in well, the enthalpy of macrocycle formation may be reduced, as with Cu^{2+} and cyclen. A second feature relates to the energy required to desolvate and reorganize the structure of the polydentate ligand. If a ligand in its lowest energy conformation already has its donor atoms in a structural arrangement which favours bonding to a metal ion, little reorganization energy is required to form the complex. The ligand can be said to be

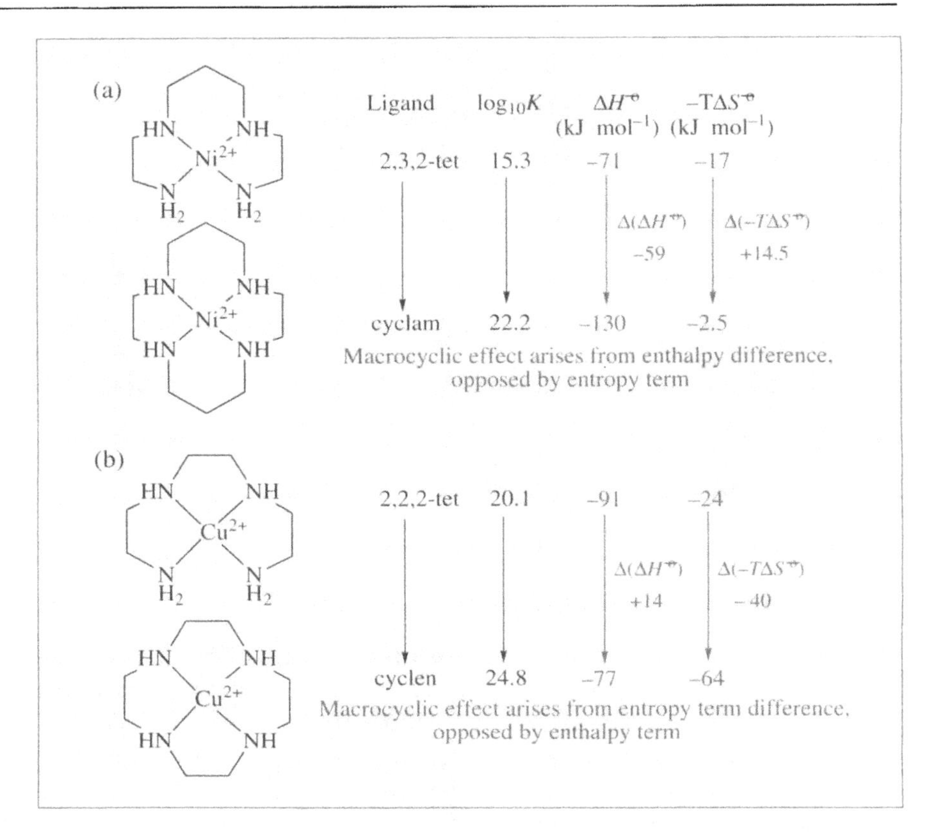

Figure 5.4 Two examples of the macrocyclic effect

preorganized. In contrast, a ligand which must adopt a structure quite different from that of its lowest energy conformation will require a reorganization energy input when forming a complex. Macrocyclic ligands such as cyclam and cyclen have four donor atoms arranged in space in a way which is predisposed to binding to a metal ion, although the nitrogen donor orbitals may need to undergo some reorientation to form the complex. In contrast, free 2,3,2-tet and 2,2,2-tet will most probably adopt extended conformations requiring more structural reorganization to form the complex (Figure 5.4). A particular example of this effect is found with the macrocyclic thioether ligand 1,4,7-trithiacyclononane (see Figure 4.4), which adopts a conformation such that the lone pairs are oriented to bind efficiently to three facial coordination sites on a metal ion. This results in the formation of particularly stable complexes, contrary to the usual observation that thioethers are rather weakly bound as ligands.

5.2.7 Steric Effects

Steric interactions between ligands can affect the stability of metal complexes and are especially important in complexes of the lanthanide ions, which are labile and do not show directional bonding. In order for a lanthanide ion complex to be stable the ligand set must effectively occupy the metal ion coordination sphere. A ligand may be assigned a solid angle factor (SAF) and the sums of the SAFs for a complex constitute a solid angle sum (SAS). Structural studies of 140 coordination compounds of lanthanide ions showed them to have a mean SAS of 0.78 ($\sigma = 0.05$). A similar study of 40 organometallic lanthanide complexes gave a mean SAS of 0.73 ($\sigma = 0.05$). A lanthanide complex with an SAS significantly lower than these norms might be expected to add ligands, or disproportionate, while higher SAS values would suggest that ligand dissociation is likely. As an example, for the complex **5.5** the SAS is calculated to be 0.768, with SAF contributions of 0.284 from each tris(pyrazolyl)borate ligand and 0.200 from the 2-formylphenolate ligand.

The **solid angle factor** of a ligand may be defined as the proportion of a sphere of radius 100 pm (1 Å) around the metal ion which is occupied by the ligand (see Figure 5.5). The solid angle sum is the sum of the solid angle factors of the ligands in the complex.

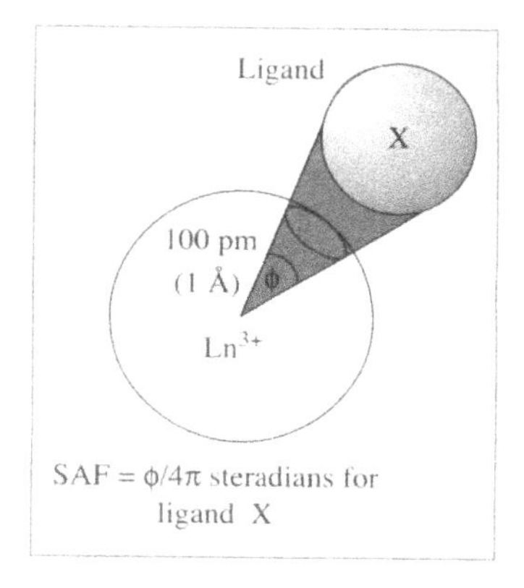

Figure 5.5 A schematic diagram showing the solid angle subtended by a ligand X and its use in the calculation of an SAF value

5.5

5.2.8 Redox Potentials

The formation of different oxidation states is an important feature of transition metal chemistry, as seen in the binary compounds described in Chapter 3. The conversion from one metal oxidation state to another in solution involves the transfer of electrons and this process is, by convention, written as a reduction (equation 5.20):

$$M^{z+} + ne^- \rightleftharpoons M^{(z-n)+} \quad (5.20)$$

The standard electrode potentials, $E^\ominus$, for such reduction reactions are related to the free energy change for the process by equation 5.3. Since some elements may exist in a number of different oxidation states, it is possible to construct electrode potential diagrams, sometimes called Latimer diagrams, relating the various oxidation states by their redox potentials. Examples are shown in Figure 5.6 for aqueous solutions of some first-row d-block metals and for some actinides in 1 mol dm^{-3} acid. In cases where the reduction involves oxide or hydroxide ions bound to

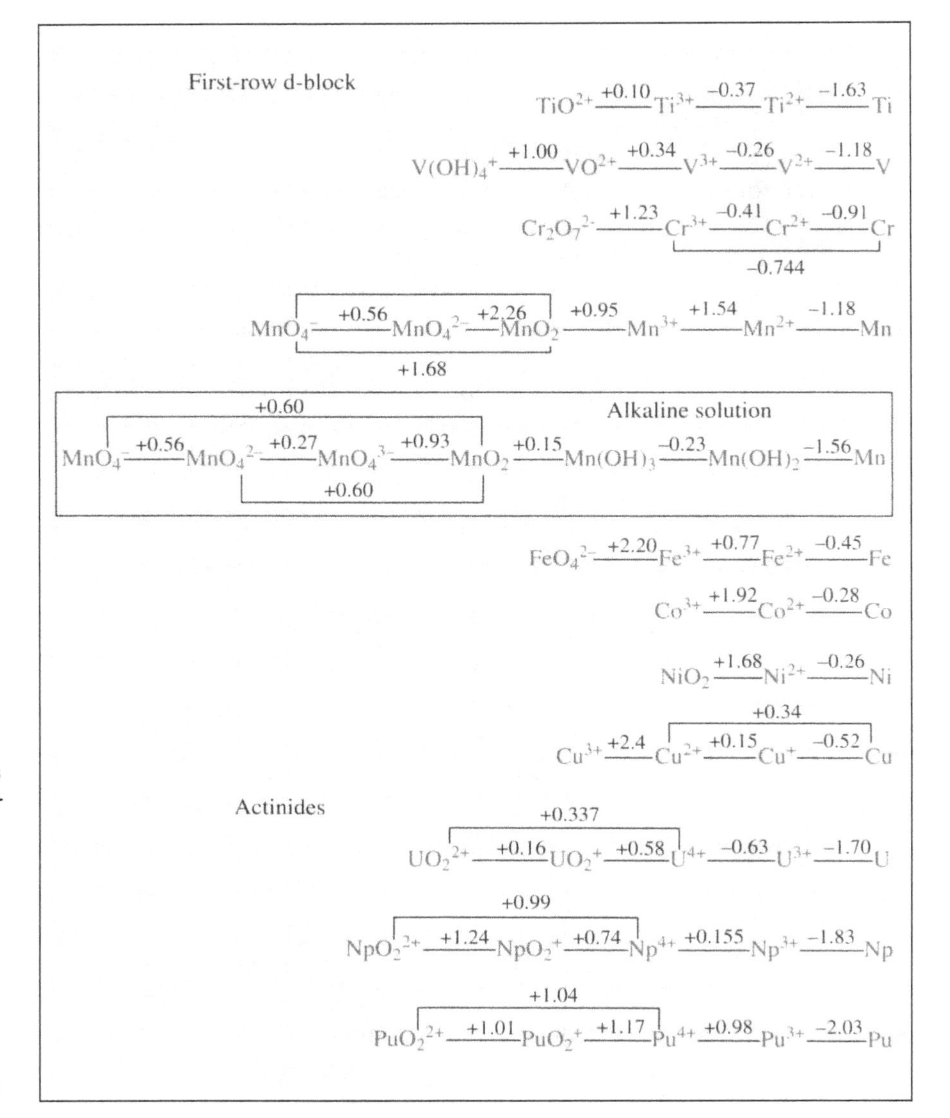

Figure 5.6 Electrode potentials from acid solutions, unless otherwise specified, for first-row d-block metal ions from *CRC Handbook of Chemistry and Physics*, 78th edn., ed. D. R. Lide, CRC Press, Boca Raton, 1997, and for selected actinides from F. A. Cotton and G. Wilkinson, *Advanced Inorganic Chemistry*, 5th edn., table 24.7, Wiley, New York, 1988. Figures are rounded to 2 decimal places

the metal, the reduction potential will be dependent upon pH. As an example, the reduction of $[MnO_4]^-$ to Mn^{2+} involves the addition of $5e^-$ and the consumption of $8H^+$ to produce $4H_2O$ from the oxide ions bound to the Mn(+7). Thus the equilibrium involves hydrogen ions, so that the free energy change, and hence $E^{\ominus}$, for reduction will vary with hydrogen ion concentration. The effect of this is illustrated in Figure 5.6 for the case of manganese, where the Latimer diagram for alkaline conditions is included.

The redox potential of a metal complex is not only dependent on the nature of the metal ion. Complexation affects the redox potentials of metal ions through the differing stability constants which arise with different ligands. The potential ($E^{\ominus}_{ML}$) at which a complex is reduced is related to the standard potential of the metal ion ($E^{\ominus}_{M}$) and the stability constants of the oxidized (β_{ox}) and reduced (β_{rd}) forms of the complex by equation 5.21 (see Box 5.3):

$$E^{\ominus}_{ML} = E^{\ominus}_{M} - (RT/nF)\ln(\beta_{ox}/\beta_{rd}) \quad (5.21)$$

Box 5.3 The Effect of Complexation on Metal Ion Redox Potential

The relationship between the electrode potential ($E^{\ominus}_{ML}$) for the reduction of a metal complex, the electrode potential of the uncomplexed metal ion ($E^{\ominus}_{ML}$), the overall stability constant for the formation of the oxidized form of the complex (β_{ox}) and the the overall stability constant for the formation of the reduced form of the complex (β_{rd}) can be derived by considering the cycle shown in Scheme 5.1:

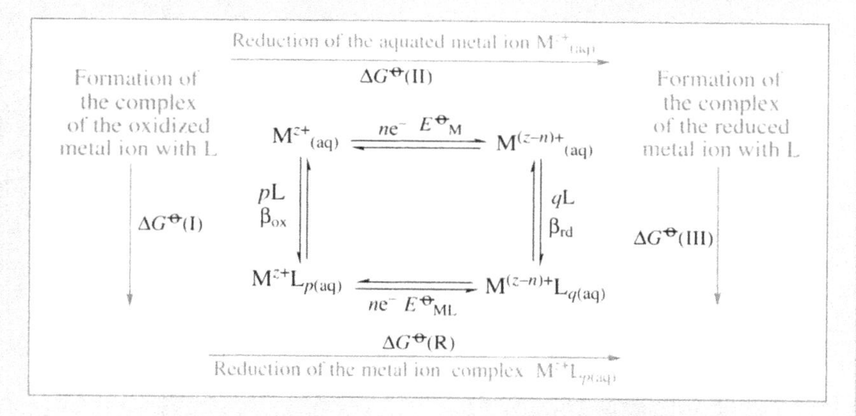

Scheme 5.1

The two pathways from M^{z+} to $M^{(z-n)+}L_q$ involve the same energy change, so that equation 5.22 can be constructed and rearranged to equation 5.23:

$$\Delta G^{\ominus}(I) + \Delta G^{\ominus}(R) = \Delta G^{\ominus}(II) + \Delta G^{\ominus}(III) \quad (5.22)$$

Hence:

$$\Delta G^{\ominus}(R) = \Delta G^{\ominus}(II) + \Delta G^{\ominus}(III) - \Delta G^{\ominus}(I) \quad (5.23)$$

Substituting for the $\Delta G^{\ominus}$ values using equations 5.1 and 5.3 gives equation 5.24:

$$-nFE^{\ominus}_{ML} = -nFE^{\ominus}_{M} + (-RT\ln\beta_{rd}) - (-RT\ln\beta_{ox}) \quad (5.24)$$

Dividing through equation 5.24 by $-nF$ gives equation 5.25:

$$E^{\text{o}}_{\text{ML}} = E^{\text{o}}_{\text{M}} + (RT/nF)\ln\beta_{\text{rd}} - (RT/nF)\ln\beta_{\text{ox}} \quad (5.25)$$

which rearranges to equation 5.26:

$$E^{\text{o}}_{\text{ML}} = E^{\text{o}}_{\text{M}} + (RT/nF)\ln(\beta_{\text{rd}}/\beta_{\text{ox}}) \quad (5.26)$$

Inverting the logarithmic term of equation 5.26 gives equation 5.21.

As an example of the application of this formula, the ratio of the stability constants for the formation of $[PdCl_4]^{2-}$ and $[PdCl_6]^{2-}$ may be obtained from electrode potential data. The difference in the reduction potentials of these two complexes is 310 mV and corresponds to the stability constant for the Pd^{4+} complex being about $10^{10.5}$ times larger than for the Pd^{2+} complex.

Worked Problem 5.3

Q Given the following redox potentials for the two-electron reductions of aqueous Pd^{4+} and aqueous $[PdCl_6]^{2-}$, calculate the ratio of the overall formation constant, β_{ox}, of $[PdCl_6]^{2-}_{(aq)}$ to that, β_{rd}, of $[PdCl_4]^{2-}$.

$$Pd^{4+}_{(aq)} + 2e^-_{(aq)} \rightleftharpoons Pd^{2+}_{(aq)} \qquad E^{\text{o}} = +1.60\ \text{V}$$

$$[PdCl_6]^{2-}_{(aq)} + 2e^-_{(aq)} \rightleftharpoons PdCl_4]^{2-}_{(aq)} + 2Cl^-_{(aq)} \qquad E^{\text{o}} = +1.29\ \text{V}$$

A Applying equation 5.22 to Scheme 5.2 allows the ratio of the formation constants for the chloride complexes of Pd^{4+} and Pd^{2+} to be determined.

$$\begin{array}{ccc} Pd^{4+}_{(aq)} & \xrightarrow{2e^-,\ +1.60\ \text{V}} & Pd^{2+}_{(aq)} \\ 6Cl^-\ \beta_{ox} \updownarrow & & \updownarrow 4Cl^-\ \beta_{rd} \\ [PdCl_6]^{2-}_{(aq)} & \xrightarrow{2e^-,\ +1.29\ \text{V}} & [PdCl_4]^{2-}_{(aq)} \end{array}$$

Scheme 5.2

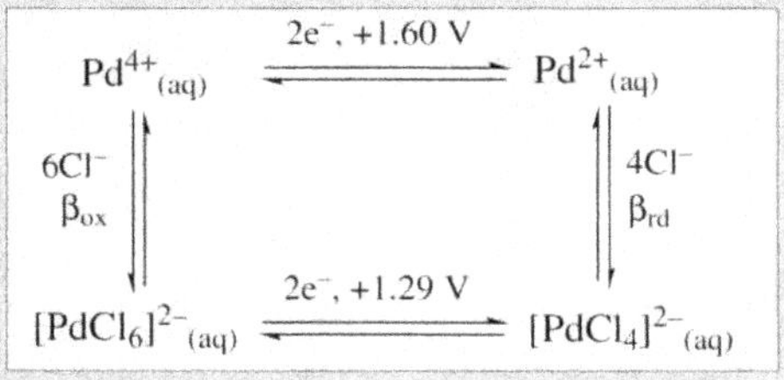

$$1.29 = 1.60 - (RT/2F)\ln(\beta_{ox}/\beta_{rd})$$

Since $2.303(RT/F) = 0.059$ V at 298 K, then:

$$1.29 - 1.60 = -(0.059/2)\log_{10}(\beta_{ox}/\beta_{rd})$$
$$0.31/0.0295 = \log_{10}(\beta_{ox}/\beta_{rd}) = 10.5$$
$$\beta_{ox}/\beta_{rd} = 10^{10.5}$$

At an intuitive level it might be expected that the redox potentials of metal complexes with polyatomic ligands might be affected by changes in ligand structure. Since the nature of the donor atoms affects the stability constant for a particular metal, this should also affect the redox potential. Similarly, the presence of electron-releasing substituents might be expected to make reduction more difficult and oxidation easier. Other factors, such as the extent of unsaturation within the ligand, may also have an effect on redox potential (Figure 5.7). The greater the degree of unsaturation and conjugation within the ligand system, the better it is at delocalizing added electrons, and the easier it is to reduce the complex.

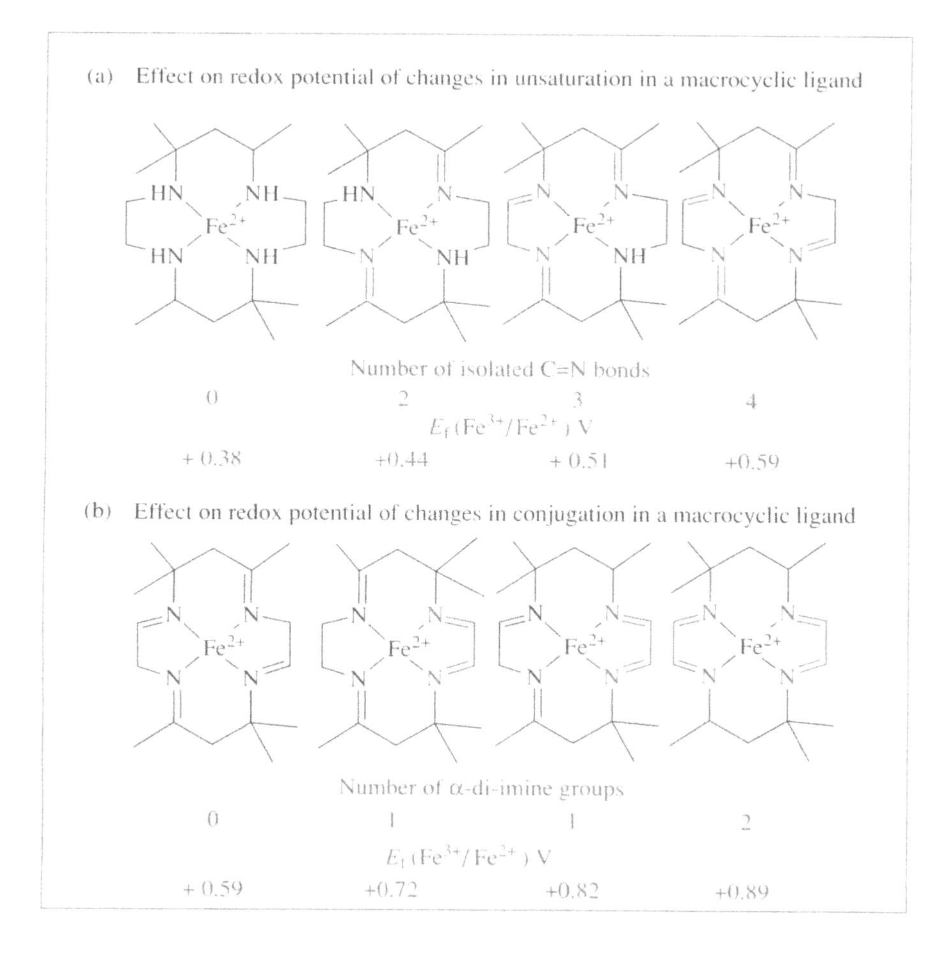

Figure 5.7 The effect of ligand structure on reduction potentials in some macrocyclic complexes of iron. (a) The effect of unsaturation; (b) the effect of increased conjugation (E_f represents the measured reduction potential under the conditions used; as these were not the standard conditions, a value of the standard potential was not obtained, but E_f can be taken as a good approximation of the standard potential)

5.3 Self-assembly and Metal Templated Reactions

5.3.1 Self-assembly

The ability of metal ions to bring together a number of ligands in an equilibrium-driven reaction can be exploited in the construction of multinuclear structures. If a rigid bifunctional ligand, with donor atoms oriented to bond to metal ions with a metal–ligand–metal angle of 180°, is combined with a metal ion which favours a square planar geometry and 90° angles at the metal, a square grid might be expected to form. An example of such a ligand–metal ion combination is provided by 4,4′-bipyridine (4,4′-bpy) and Pd^{2+}. If two *cis* coordination sites on the Pd^{2+} ion are blocked by a chelating ligand such as ethane-1,2-diamine, a tetranuclear complex with a square structure can form (Figure 5.8). Since Pd^{2+} exchanges ligands rapidly it can quickly explore all possible structural arrangements in an equilibrium-driven reaction. In practice, the tetranuclear square formed from Pd^{2+} and 4,4′-bpy is of low solubility, so it is precipitated from aqueous solution. Through Le Chatelier's principle, this leads to high yields of the cyclic tetranuclear product at the expense of the other oligomers which may form. This reaction is an example of **molecular self-assembly** and, since its discovery, has been shown to lead to a diverse set of polynuclear structural types, a further example of which is shown in Figure 5.9.

The term **molecular self-assembly** refers to a process in which a number of components bind together at equililbrium to form a particular product in high yield. At its simplest level this could apply to the formation of a metal complex such as $[Co(NH_3)_6]^{2+}$ from Co^{2+} and six NH_3 molecules, where seven entities combine into one. However, the term is more usually applied to the formation of more elaborate structures, such as the polynuclear complexes shown in Figures 5.8 and 5.9.

5.3.2 Metal Templated Reactions

The ability of metal ions to bring together a number of ligands in a complex provides a means of promoting synthetic processes which involve reactions between the coordinated ligands. In such cases the metal ion acts as a template, orienting and perhaps activating the reactants. A metal templated reaction occurs when the precursors to a ligand only assemble to form the ligand in the presence of a metal ion. In the absence

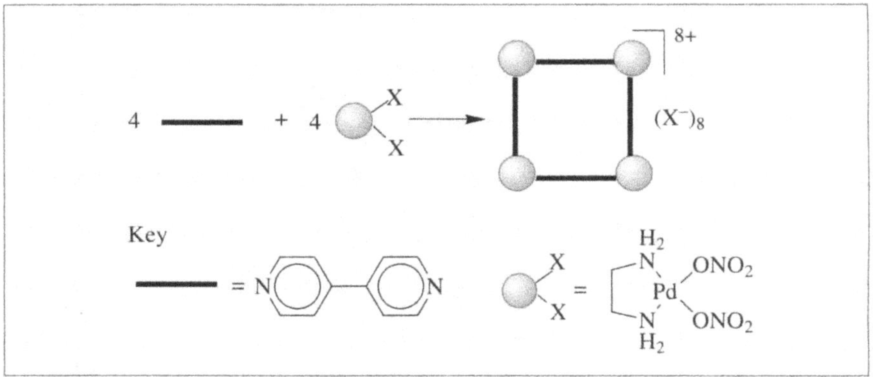

Figure 5.8 Self-assembly of a tetranuclear molecule

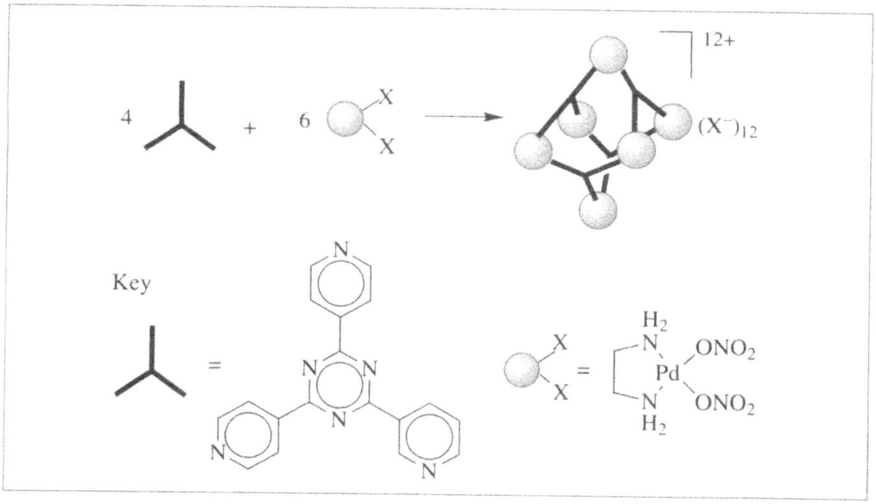

Figure 5.9 Self-assembly of a hexanuclear molecule

of a metal ion the reaction either does not proceed or produces a different product. A good example of this is provided by the self-reaction of phthalonitrile, 1,2-$(CN)_2C_6H_4$, which, in the presence of metal halides such as $CuCl_2$, gives the macrocyclic phthalocyanine ligand as its metal complex (Figure 5.10a). In the absence of a metal, intractable polymeric materials are formed. Similarly, the condensation of 2-aminobenzaldehyde proceeds in the presence of metal ions such as Ni^{2+} to form a macrocycle as its metal complex (Figure 5.10b).

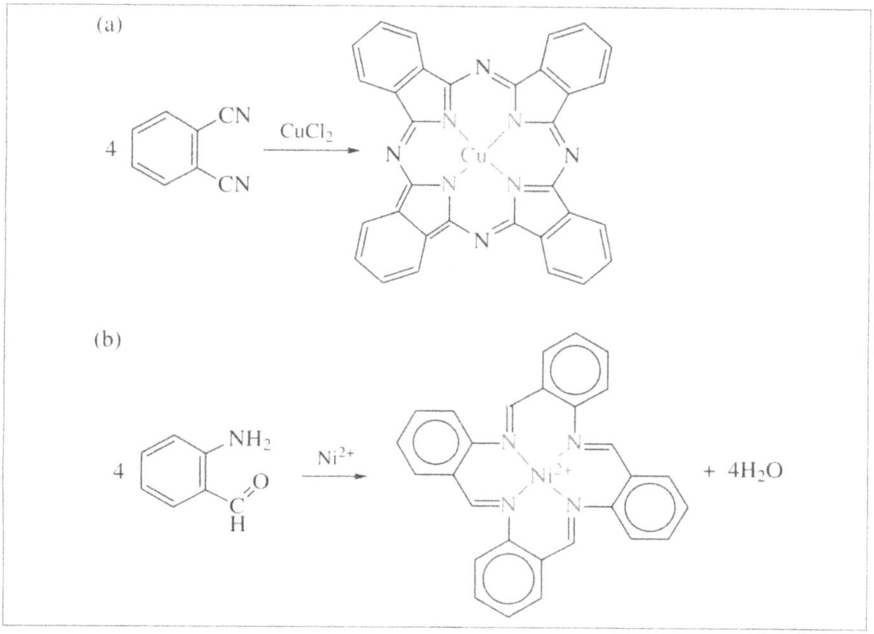

Figure 5.10 Examples of metal templated syntheses

Another example of metal templated ligand assembly is provided by the reaction of propan-2-one with ethane-1,2-diamine. In the absence of a metal ion these reagents will react to form the *trans* form of a macrocyclic proligand (Figure 5.11a), but the *cis* form of the macrocycle only forms in the presence of Ni^{2+} *via* a template reaction (Figure 5.11b). A possible mechanism for the macrocycle formation reaction involves the base-induced condensation of a methyl group with a nearby imine group (Figure 5.11b).

Two primary types of template reaction have been identified and these demonstrate equilibrium template (see Box 5.4) and kinetic template effects. An example of an equilibrium template effect is provided by the reaction between butane-2,3-dione and 2-aminoethanethiol which produces, as its major product, a heterocycle in equilibrium with smaller amounts of the acyclic imine (Scheme 5.3). However, in the presence of Ni^{2+} ions the acyclic product forms in high yield as its nickel complex.

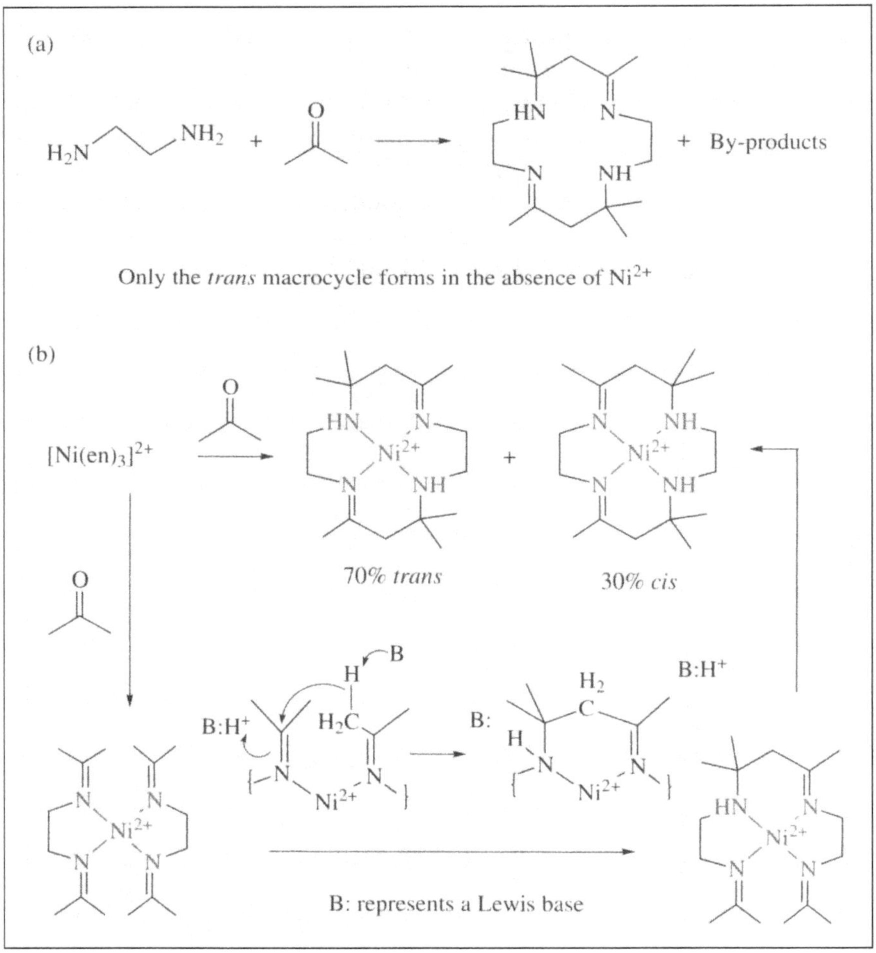

(a)

H_2N NH_2 + O → HN N N NH + By-products

Only the *trans* macrocycle forms in the absence of Ni^{2+}

(b)

$[Ni(en)_3]^{2+}$

70% *trans*

30% *cis*

B:H⁺

B: represents a Lewis base

Figure 5.11 A metal templated reaction: (a) the reaction in the absence of Ni^{2+} ions; (b) the reaction in the presence of Ni^{2+} ions

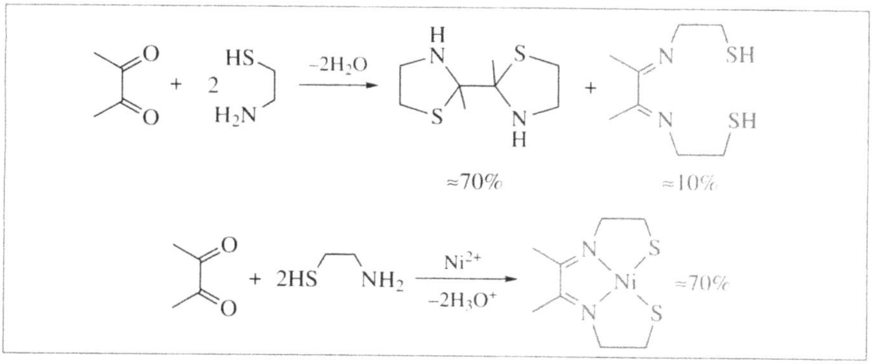

Scheme 5.3

Box 5.4 The Equilibrium Template Reaction

In a reaction to produce two or more products, complexation to a metal ion selectively stabilizes one component of the equilibrium mixture, favouring its production as the metal complex (Figure 5.12). This has the effect of shifting the equilibrium distribution of products so that instead of B being favoured over A, A becomes favoured over B in the form of its complex [MA].

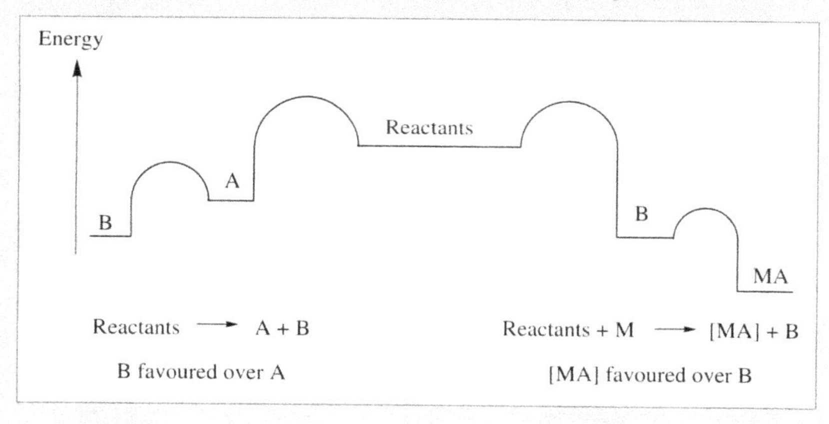

Figure 5.12 The equilibrium template reaction

An example of a kinetic template effect is provided by the reaction between this complex and 1,2-bis(bromomethyl)benzene. Complexation orients the sulfur atoms of the quadridentate N_2S_2 ligand to favour ring formation in a two-step reaction with 1,2-$(BrCH_2)_2C_6H_4$, as shown in Scheme 5.4. In the absence of the metal ion, reaction of the free ligand with 1,2-$(BrCH_2)_2C_6H_4$ gives polymeric materials.

In a **kinetic template reaction**, one or more reactants are bonded to a metal ion to form a complex which does not easily undergo ligand exchange reactions. This orients the reactants in a way which favours a particular stereoselective multistep reaction sequence, leading to the formation of a particular product as its metal complex.

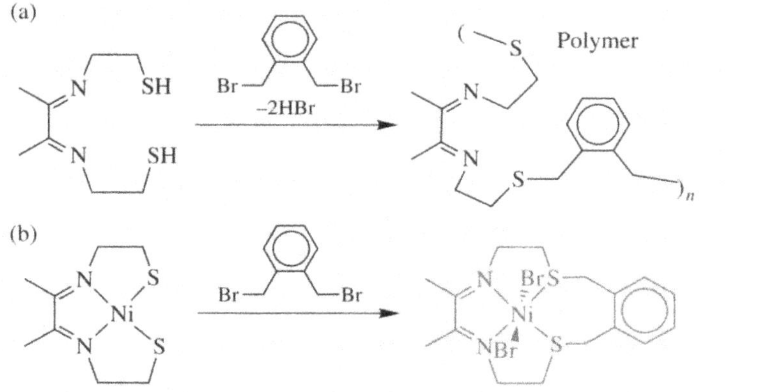

Scheme 5.4

Summary of Key Points

1. *The driving force for the formation of a coordination compound* may be expressed in the form of an equilibrium constant known as a stability constant or formation constant.

2. *The stability constant of a metal complex* depends upon the nature of the ligands and the nature of the metal ion. Normally hard metals bind more strongly to hard ligand donor atoms.

3. *Ligands which bind to metals through more than one donor atom* show relatively high stability constants. Both enthalpy and entropy contribute to the magnitude of the stability constant and the balance between these contributions varies with ligand structure.

4. *Redox potentials for metal ions* are strongly influenced by the presence of coordinated ligands.

5. *Metal ions can act as templates*, directing reactions between the precursors of polydentate ligands or the formation of polynuclear structures.

Problems

[Assume the following numerical values: $2.303\log(x) = \ln(x)$; $R = 8.314 \text{ J K}^{-1} \text{ mol}^{-1}$; $2.303RT/F = 0.059$ V; $2.303RT = 5.706 \text{ kJ mol}^{-1}$ at 298 K].

5.1. Calculate the equilibrium constant for the reaction of $\{AgF\}_{(aq)}$ with $\{Ca(NH_3)\}^{2+}_{(aq)}$ according to equation 5.27:

$$\{AgF\}_{(aq)} + \{Ca(NH_3)\}^{2+}_{(aq)} \rightleftharpoons \{Ag(NH_3)\}^{+}_{(aq)} + \{CaF\}^{+}_{(aq)} \quad (5.27)$$

The $\log_{10}K$ values for equations 5.28 and 5.29 are:

$$NH_{3(aq)} + \{CaF\}^{+}_{(aq)} \rightleftharpoons \{Ca(NH_3)\}^{2+}_{(aq)} + F^{-}_{(aq)} \quad \log_{10}K = -0.7 \quad (5.28)$$

$$NH_{3(aq)} + \{AgF\}_{(aq)} \rightleftharpoons \{Ag(NH_3)\}^{+}_{(aq)} + F^{-}_{(aq)} \quad \log_{10}K = 2.96 \quad (5.29)$$

5.2. (i) Explain what is meant by the terms *hard* and *soft* when applied to metal ions.
(ii) In the electron transfer protein cytochrome *c* the iron ion is coordinated to six nitrogen atoms, but in cytochrome *b* the iron ion is coordinated to one sulfur and five nitrogen atoms. How would you expect this to affect the potential of the Fe^{3+} to Fe^{2+} reduction process and on what basis?

5.3. Calculate the overall stability constant β_{rd} for the complexation of Fe^{2+} by cyanide at 298 K according to equation 5.30, using the data given for equations 5.31–5.33.

$$Fe^{2+}_{(aq)} + 6CN^{-}_{(aq)} \rightleftharpoons [Fe(CN)_6]^{4-}_{(aq)} \quad (5.30)$$

$$Fe^{3+}_{(aq)} + e^{-}_{(aq)} \rightleftharpoons Fe^{2+}_{(aq)} \quad E^{\ominus} = +0.77\ V \quad (5.31)$$

$$[Fe(CN)_6]^{3-}_{(aq)} + e^{-}_{(aq)} \rightleftharpoons [Fe(CN)_6]^{4-}_{(aq)} \quad E^{\ominus} = +0.36\ V \quad (5.32)$$

$$Fe^{3+}_{(aq)} + 6CN^{-}_{(aq)} \rightleftharpoons [Fe(CN)_6]^{3-}_{(aq)} \quad \log_{10}\beta_{ox} = 31 \quad (5.33)$$

5.4. Calculate the overall stability constant β for the formation of $[Au(CN)_2]^{-}$ according to equation 5.34, using the data given for equations 5.35 and 5.36:

$$Au^{+}_{(aq)} + 2CN^{-}_{(aq)} \rightleftharpoons [Au(CN)_2]^{-}_{(aq)} \quad (5.34)$$

$$Au^{+}_{(aq)} + e^{-}_{(aq)} \rightleftharpoons Au \quad E^{\ominus} = +1.69\ V \quad (5.35)$$

$$[Au(CN)_2]^{-}_{(aq)} + e^{-}_{(aq)} \rightleftharpoons Au + 2CN^{-}_{(aq)} \quad E^{\ominus} = -0.60\ V \quad (5.36)$$

5.5. (i) Explain what is meant by the term *macrocyclic effect*.
(ii) Calculate the stability constant for the formation of $\{Ni(cyclam)\}^{2+}{}_{(aq)}$ in water at 298 K according to equation 5.37, given the data for equations 5.38 and 5.39. Comment on the origins of the macrocyclic effect in this case (2,3,2-tet and cyclam are defined in Figures 4.3 and 4.4, respectively).

$$Ni^{2+}{}_{(aq)} + cyclam_{(aq)} \rightleftharpoons \{Ni(cyclam)\}^{2+}{}_{(aq)} \quad (5.37)$$

$$Ni^{2+}{}_{(aq)} + 2,3,2\text{-}tet_{(aq)} \rightleftharpoons \{Ni(2,3,2\text{-}tet)\}^{2+}{}_{(aq)} \quad (5.38)$$

$\Delta H^{\ominus} = -71$ kJ mol^{-1}, $\Delta S^{\ominus} = 55$ J K^{-1} mol^{-1}

$$\{Ni(2,3,2\text{-}tet)\}^{2+}{}_{(aq)} + cyclam_{(aq)} \rightleftharpoons \{Ni(cyclam)\}^{2+}{}_{(aq)} + 2,3,2\text{-}tet_{(aq)} \quad (5.39)$$

$\Delta H^{\ominus} = -59$ kJ mol^{-1}, $\Delta S^{\ominus} = -47$ J K^{-1} mol^{-1}

5.6. Explain the trends in the magnitudes of the $\log_{10}K_1$ values for the formation of the mono pentane-2,4-dionate complexes of lanthanide and actinide ions shown in Table 5.2. Consider the variation across the lanthanide series, the variation across the actinide series, and the differences between the two series.

Table 5.2 Values of $\log_{10}K_1$ for the formation of $\{Ln(acac)\}^{2+}{}_{(aq)}$, $\{An(acac)\}^{3+}{}_{(aq)}$ and $\{AnO_2(acac)\}^{z+}{}_{(aq)}$ (An = U, z = 1; An = Np, z = 0)

Metal ion	Y^{3+}	La^{3+}	Ce^{3+}	Pr^{3+}	Nd^{3+}	Sm^{3+}	Eu^{3+}	Gd^{3+}	Tb^{3+}	Dy^{3+}	Ho^{3+}	Er^{3+}	Tm^{3+}	Yb^{3+}	Lu^{3+}
$\log_{10}K_1$	5.89	4.94	5.15	5.35	5.36	5.67	5.94	5.90	6.02	6.06	6.07	6.08	6.14	6.18	6.15
Metal ion	Th^{4+}	U^{4+}	Pu^{4+}	UO_2^{2+}	NpO_2^{+}										
$\log_{10}K_1$	7.7	8.6	10.5	6.8	4.08										

5.7. Define the terms 'equilibrium template reaction' and 'kinetic template reaction' as applied to reactions involving transition metal ions, giving an example of each.

6 Bonding in Transition Metal Complexes

Aims

After reading this chapter you should have an understanding of bonding in transition metal complexes and a knowledge of:

- The crystal field theory as applied to octahedral, tetragonal, tetrahedral and square planar metal complexes
- The meaning of the terms low or high spin as applied to transition metal complexes and the terms strong or weak field as applied to ligands
- The meaning of the term crystal field stabilization energy and how it is calculated
- The Jahn–Teller effect and its implications for the structures of transition metal complexes
- The molecular orbital theory as applied to octahedral and tetrahedral metal complexes.

6.1 Introduction

In Chapter 3 the lattice enthalpies of some binary first-row transition metal dichlorides were found to deviate from a simple variation with atomic number (Figure 3.3). Although ions with d^0, d^5 and d^{10} electron configurations fit a simple curve, ions with other electron configurations show larger than expected lattice enthalpies. The hydration enthalpies of first-row d-block M^{2+} ions show similar deviations (Figure 5.1) but, in contrast, the hydration enthalpies of the lanthanide Ln^{3+} ions follow a fairly smooth curve, corresponding with their steadily decreasing ionic radii (Figure 5.2). This difference in behaviour suggests that some additional bonding interaction is present in the compounds of certain first-row d-block metals compared to others and to the lanthanide ions. In fact the plots of hydration energy against atomic number reflect the differing extents to which the valence shell orbitals of d-block and f-block metal ions interact with the ligands, water in this case. The 4f orbitals of the lanthanides are core-like and are little affected by the presence of the water ligands. However, the 3d orbitals of the first-row d-block metals are less core-like and interact more strongly with the ligands. It is this interaction, combined with the varying number of valence shell electrons, which leads to the deviations from a simple curve in the case of the d-block metal ions. The earliest theory explaining the nature of this interaction is known as the crystal field theory (CFT). This ionic bonding model was developed to explain the properties of metal ions in crystal lattices. The CFT model was further developed to allow for covalent contributions to metal–ligand bonding in the ligand field theory (LFT). The CFT and LFT models explain many of the basic features of transition metal compounds, including some of their their magnetic and spectroscopic properties, but they have some limitations. A molecular orbital (MO) theory approach provides a more universal treatment of bonding. It naturally incorporates covalency, and so offers a more generally applicable model which can readily embrace organometallic compounds. However, despite its limitations, CFT is simple to apply and provides a useful qualitative model which is still widely used. In this chapter a brief account of the CFT and MO models is presented.

6.2 The Crystal Field Model

6.2.1 Octahedral Complexes

The CFT is an electrostatic model which considers the effect of the electric field due to the ligand electronic charge on the energies of electrons in the various d orbitals of the metal ion. The radial charge distribution functions of the d orbitals are represented schematically in Figure 6.1 and illustrate, for each d orbital, the region of space occupied by an electron in that orbital. These show that the d_{z^2} and $d_{x^2-y^2}$ orbitals are directed along the Cartesian axes, and the d_{xy}, d_{xz} and d_{yz} orbitals are directed between the axes. If a metal ion is removed from a vacuum into the electrostatic field created by a spherical shell of electron density equivalent to that of the six ligands in an octahedral complex, electrons in the d orbitals will move to higher energy because of the repulsive force produced by the surrounding shell of electronic charge. In an energy level diagram the energies of the five degenerate d orbitals can be represented by five lines as shown in Figure 6.2. If the spherical charge is then redistributed to six equivalent points at the vertices of an octahedron, the degeneracy of the five d orbitals is partly removed by the electrostatic field. This happens because electrons in the d_{z^2} and $d_{x^2-y^2}$ orbitals

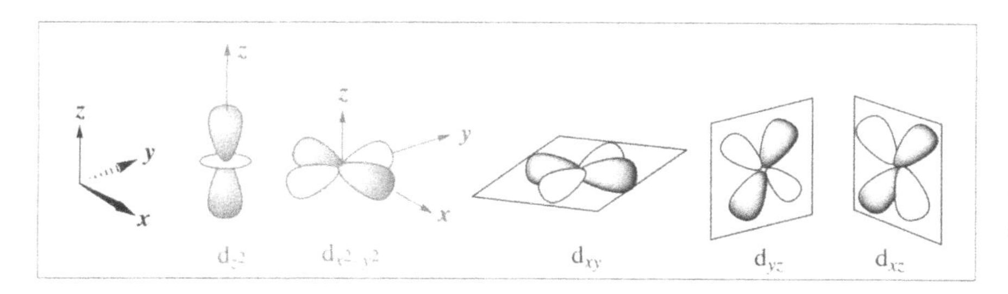

Figure 6.1 A schematic representation of the electron density distribution in d orbitals

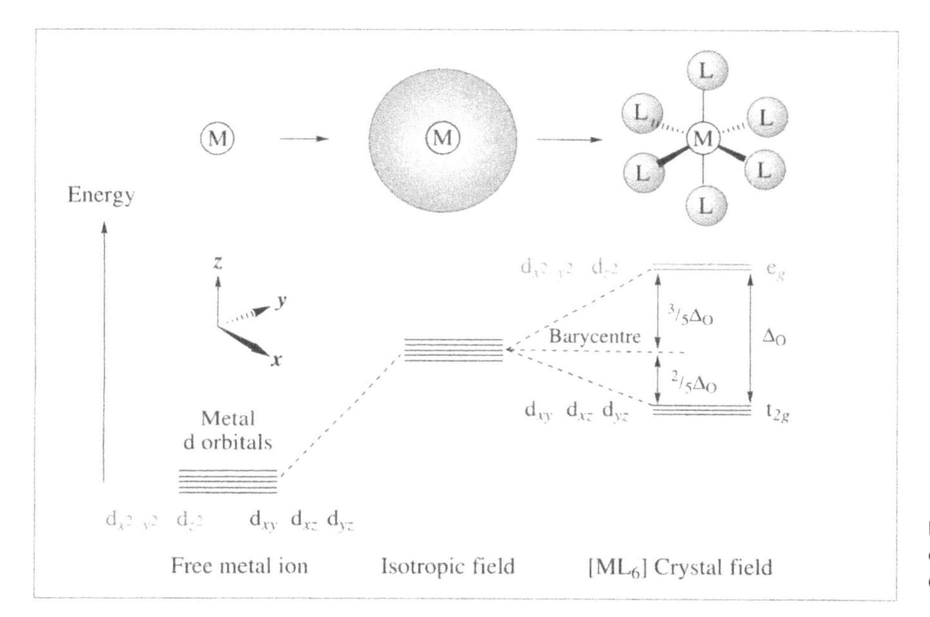

Figure 6.2 The effect of an octahedral crystal field on the energies of d orbitals.

will, on average, spend more time closer to the six regions of ligand electron density than those in the d_{xy}, d_{xz} and d_{yz} orbitals. Thus, in an energy level diagram, the d_{z^2} and $d_{x^2-y^2}$ orbitals would be placed higher in energy than the d_{xy}, d_{xz} and d_{yz} orbitals. The difference in energies between the two sets of orbitals is known as the crystal field splitting or ligand field splitting. This splitting is represented by the crystal field splitting parameter Δ, also sometimes called $10Dq$ where $10Dq = \Delta$. The more specific symbol Δ_O may be used in the particular case of an octahedral complex. Similarly, the symbol Δ_T may be used to denote the value for a tetrahedral complex.

The two sets of orbitals are given symmetry labels (Box 6.1), e_g for the two degenerate d_{z^2} and $d_{x^2-y^2}$ orbitals and t_{2g} for the three degenerate d_{xy}, d_{xz} and d_{yz} orbitals. Since rearranging the charge within the spherical shell does not produce any change in the total energy of the system, the barycentre, or energy centre of gravity, of the set of d orbitals must remain the same. Thus the energy of the two e_g orbitals will increase by

($\frac{3}{5}$)Δ_O and that of the three t_{2g} orbitals will decrease by by ($\frac{2}{5}$)Δ_O. This means that, in a filled subshell d^{10} ion, four electrons will move to higher energy by ($\frac{3}{5}$)Δ_O each and six electrons will move to lower energy by ($\frac{2}{5}$)Δ_O each. The result for the d^{10} closed subshell system is no change in energy. A similar result would be obtained for a d^0 or and a d^5 electron configuration, so that the energies of Sc^{3+}, Mn^{2+}, Fe^{3+} and Zn^{2+} ions would not be affected by the field of the ligands. However, in a d^3 ion, such as Cr^{3+}, three electrons would occupy the lower energy t_{2g} orbital, so that the energy of the ion in an octahedral crystal field would be less than in an equivalent spherical field by 3 × ($\frac{2}{5}$)Δ_O, *i.e.* an energy change of –($\frac{6}{5}$)Δ_O. This additional stabilization of the complex, compared to an equivalent closed subshell metal ion, is known as the crystal field stabilization energy (CFSE) and depends upon the d electron configuration of the metal ion, as shown in Figure 6.3. In the case of an octahedral metal complex with the electron configuration $t_{2g}{}^{x}e_g{}^{y}$ and a crystal field splitting Δ_O, the CFSE is given by equation 6.1:

For the **crystal field stabilization energy**, following thermodynamic convention the energies of electrons entering the e_g orbitals, at higher energy than the barycentre, are counted as positive whilst the energies of those entering the t_{2g} orbitals, at lower energy than the barycentre, are assigned a negative energy. When the sum of the energies of electrons in the orbitals is calculated, a negative number will result except in the cases of $t_{2g}{}^{0}e_g{}^{0}$, $t_{2g}{}^{3}e_g{}^{2}$ and $t_{2g}{}^{6}e_g{}^{4}$ electron configurations, where there will be zero energy change compared to the situation with an isotropic field.

$$\text{CFSE} = \{y(\tfrac{3}{5}) - x(\tfrac{2}{5})\}\Delta_O \qquad (6.1)$$

Box 6.1 Crystal Field Theory and Symmetry Labels

Group theory provides the symmetry basis for MO theory, and the symmetry labels e_g and t_{2g} have a particular significance which those who are familiar with chemical applications of group theory will appreciate. Those unfamiliar with group threory should simply accept these symbols as labels used to denote the symmetries of particular sets of orbitals. In the symmetry of an octahedron the directions *x*, *y* and *z* are equivalent and cannot be distinguished one from another. In terms of symmetry, *x*, *y* and *z* must be treated as a set of three inseparable entities. Similarly, in an octahedral environment, the d_{xy}, d_{xz} and d_{yz} orbitals cannot be distinguished by symmetry and have the same energy. They are said to be degenerate, and are given the symmetry label t_{2g} in which the t signifies that there are three degenerate orbitals and the subscript $_g$ stands for the term 'gerade', meaning that they are symmetrical with respect to the inversion through the centre of the octahedron (if you invert any of the orbitals shown in Figure 6.1 through their centre, they look the same after inversion as before). The d_{z^2} and $d_{x^2-y^2}$ orbitals form a degenerate pair, and are given the symmetry label e_g in which e signifies the presence of two orbitals of the same energy which cannot be distinguished by symmetry. The three p orbitals in the valence shell of a metal ion are also degenerate in an octahedral field, and will have a symmetry label containing t. Unlike d orbitals, p orbitals change phase on inversion through a centre of symmetry. This is denoted by the subscript $_u$ (ungerade), so that the three degenerate p orbitals are given the symmetry label t_{1u}. The valence shell s orbital is spherically symmetric and has the symmetry label a_{1g} in an octahedral environment.

This ionic model provides a good basis for explaining anomalies in the variation of lattice energies and hydration energies across the first-row d-block for octahedral metal ions. As electrons are added to the t_{2g} orbitals going from d^0 ions to d^3 ions, the CFSE increases from zero to

Figure 6.3 The electron configurations of d-block ions in an octahedral crystal field (the terms high spin and low spin are defined on page 102)

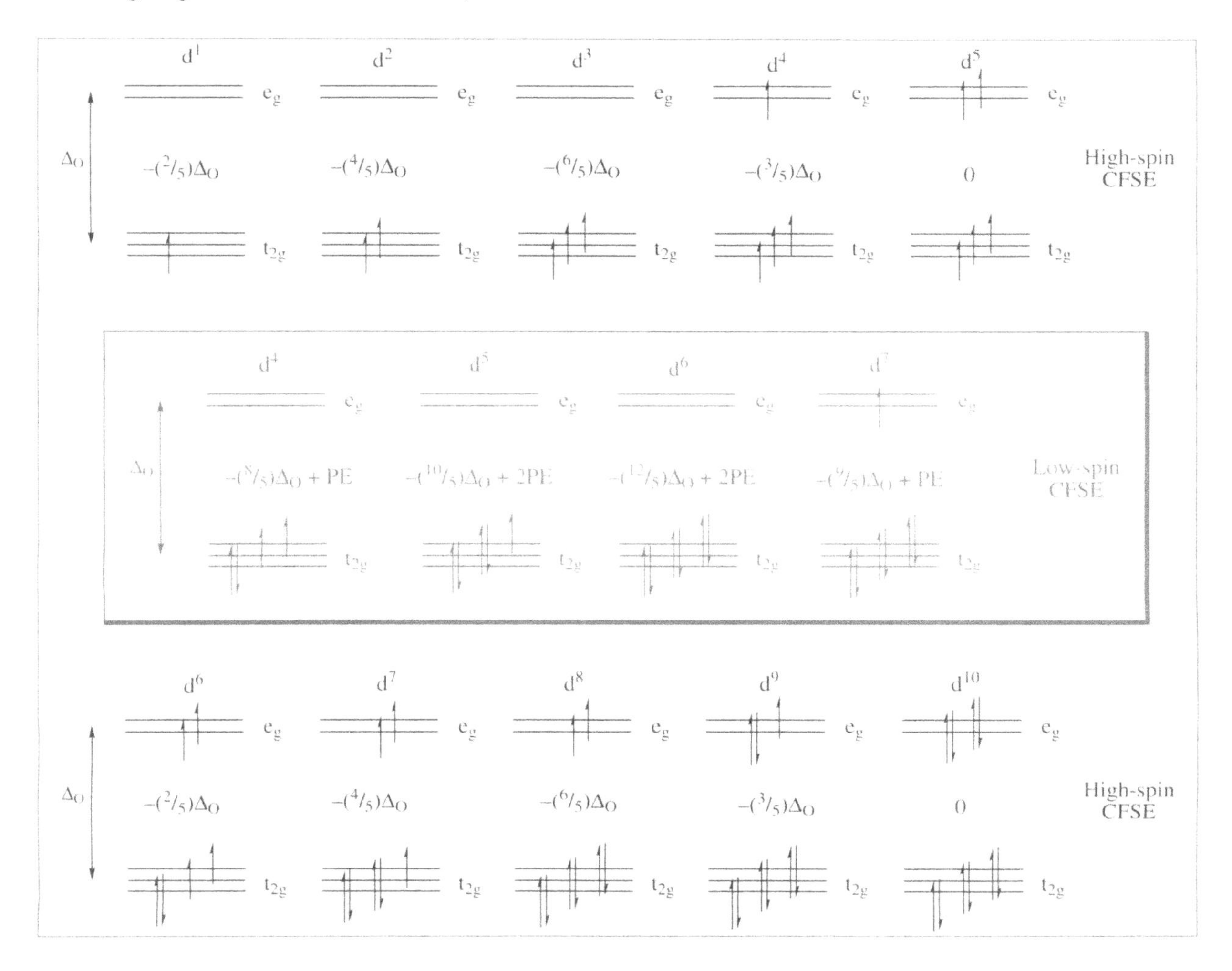

$(-\frac{6}{5})\Delta_O$, leading to a larger hydration or lattice energy. However, as electrons are added to the e_g orbital this effect is reduced, then cancelled out, with CFSE values of $(-\frac{3}{5})\Delta_O$ for d^4 and zero for d^5. Between d^5 and d^{10} this pattern is repeated. When this pattern of changing CFSE values is superimposed on a lattice or hydration energy which is steadily increasing in magnitude with increasing atomic number and decreasing atomic radius, the general form of the plots in Figures 3.3 and 5.1 emerges.

Worked Problem 6.1

Q Assuming that all have a similar structure, with metal ions in octahedral sites, explain the variation in lattice energies of the first-row d-block metal ions shown in Figure 3.3.

A The general trend to more exothermic values with increasing atomic number is attributable to the decrease in ionic radius across the period because, as the anion–cation separation becomes smaller, the lattice enthalpy increases (equation 3.3). Superimposed on this trend is the effect of CFSE values. These are small in comparison to the overall magnitude of $\Delta H^{\ominus}_U$, but nonetheless have a significant effect. The 'double dip' in the plot may be accounted for in terms of the variation in high-spin CFSE values across the first-row d-block, as shown in Figure 6.3. This shows respective CFSE contributions to $\Delta H^{\ominus}_U$ of $-(\frac{4}{5})\Delta_O$ for Ti^{2+}, $-(\frac{6}{5})\Delta_O$ for V^{2+}, $-(\frac{3}{5})\Delta_O$ for Cr^{2+}, 0 for Mn^{2+}, $-(\frac{2}{5})\Delta_O$ for Fe^{2+}, $-(\frac{4}{5})\Delta_O$ for Co^{2+}, $-(\frac{6}{5})\Delta_O$ for Ni^{2+}, $-(\frac{3}{5})\Delta_O$ for Cu^{2+} and 0 for Zn^{2+}. When superimposed on the trend due to decreasing ionic radius, this produces the double dip in the plot.

In the cases considered so far it has been assumed that electrons added beyond d^3 will occupy the e_g orbitals in preference to pairing up with electrons in the t_{2g} orbitals. This is not unreasonable, since there is an energy penalty, the pairing energy (PE, see page 25), associated with bringing electrons close together to form a pair in an orbital. However, if the interaction between the metal ion and the ligands is sufficiently large so that $\Delta_O > PE$, it becomes more energetically favourable for the fourth electron added to occupy a t_{2g} orbital. Ligands which produce this effect are known as strong-field ligands ($\Delta_O > PE$) and the complexes they form are called low-spin complexes. Ligands for which $\Delta_O <$ PE are known as weak-field ligands and form high-spin complexes. When calculating the CFSE of a strong-field low-spin complex it is necessary to include the pairing energies associated with the additional electron pairs which must form (Box 6.2). In such cases, equation 6.1 must be modified to equation 6.2:

$$(\text{CFSE})_L = \{y(3/5) - x(2/5)\}\Delta_O + \Pi(\text{PE}) \qquad (6.2)$$

in which Π represents the number of additional electron pairs and $(\text{CFSE})_L$ signifies a value of CFSE for a low-spin case.

Box 6.2 Pairing Energy and Low-spin Complexes

When considering pairing energies, it is important to note that it is not necessarily the total number of electron pairs which should be considered but the number of new electron pairs created compared to the high-spin case, as illustrated in Figure 6.4. Pairing energies (PEs) for first-row d-block metal ions vary with the metal and the charge (*e.g.* for Fe^{2+} and Fe^{3+}, respective PE values are 211 and 359 kJ mol^{-1}; for Cr^{2+} and Mn^{2+}, respective PE values are 281 and 305 kJ mol^{-1}). A major contributor to the PE is the difference in exchange energy (Box 2.1) between the high-spin and low-spin states. As an example, the exchange energy in a high-spin d^4 configuration ($t_{2g}^3e_g^1$) with four parallel electron spins is $6K^x$. This may be compared with a low-spin d^4 configuration ($t_{2g}^4e_g^0$) with three parallel electron spins and the fourth spin reversed to form a pair. This has an exchange energy of only $3K^x$. Converting from the high-spin to the low-spin case produces a change in the CFSE of $-\Delta_O$ as one electron is transferred from e_g to t_{2g}. However, this must be set against the change of $+3K^x$ in exchange energy.

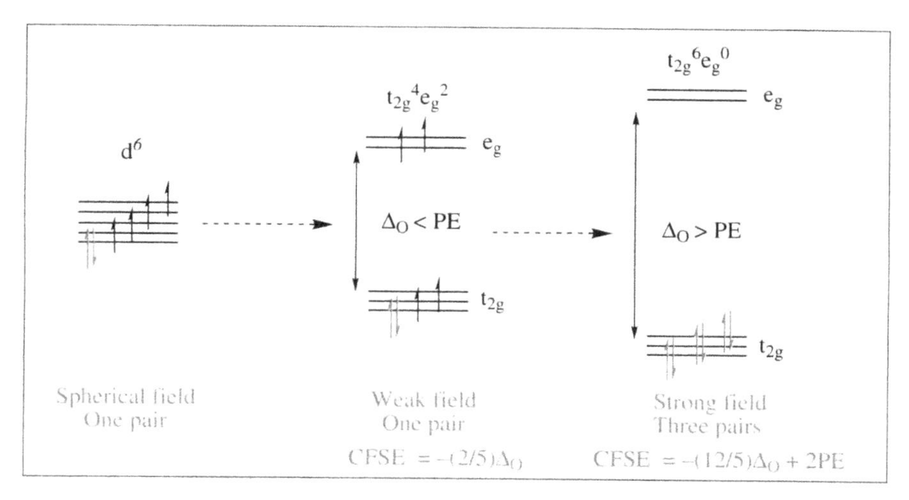

Figure 6.4 High- and low-spin configurations in an octahedral crystal field

Ionic Radii

In Chapter 2 it was seen that there is a general decrease in ionic radius with increasing atomic number, or nuclear charge, across both the d- and f-blocks. In the case of the lanthanide Ln^{3+} ions this decrease follows a uniform curve. However, detailed measurements of ionic radii in octahedral transition metal oxides and fluorides show that the trend is not uniform for d-block metal ions. An ion in a high-spin complex will show a different radius from that found in a low-spin complex (Figure 6.5a). The CFT model can be used to explain these observations. Following the argument used earlier, electrons in the e_g orbitals of an octahedral complex are closer in proximity to the ligands than those in the t_{2g} orbitals. As a consequence, ligands approaching along the Cartesian axes of a metal ion in which the e_g orbitals are occupied will encounter metal electron density earlier than would have been the case if only t_{2g} were occupied (Figure 6.5b). Thus the metal ion will appear to have a larger radius, so that the pattern in radii seen in Figure 6.5a

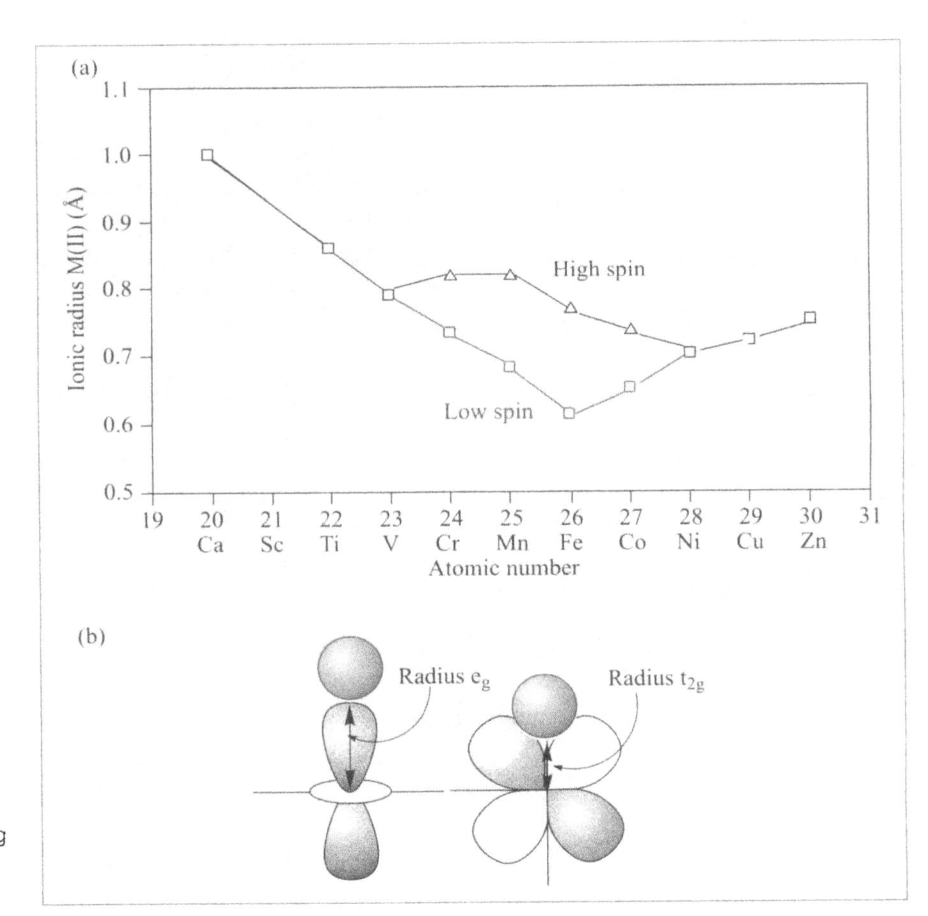

Figure 6.5 (a) The variation in ionic radii across the first-row d-block for M^{2+} ions. (b) The differing interactions between a ligating atom σ orbital and the e_g or t_{2g} orbitals

can be correlated with the high-spin metal ion electron configurations shown in Figure 6.3. The first three electrons in a high- or low-spin complex enter t_{2g}, so, between d^1 and d^3, metal ions follow a trend of decreasing ionic radius with increasing atomic number. However, when the fourth and fifth electrons are added in a high-spin complex they enter e_g, halting the decrease in radius. Subsequently, the sixth, seventh and eighth electrons go into t_{2g} so the radii decrease again slightly but, as the last two electrons are added to e_g, for the d^9 and d^{10} configurations the radii increase again. In a low-spin complex, electrons continue to enter t_{2g} until a d^6 configuration is reached, so the radii continue to decline beyond d^3 and only begin to increase from d^7 onwards as the e_g orbitals are occupied. The crystal field model also provides a basis for predicting the number of unpaired electrons in a metal ion in a complex, depending upon whether it is high or low spin as shown in Figure 6.3.

Worked Problem 6.2

Q Using the $10Dq$ values below estimated from spectroscopic measurements, calculate the crystal field stabilization energy of $[Fe(NH_3)_6]^{3+}$ in kJ mol^{-1} ($10Dq$ = 20,000 cm^{-1}; assume a pairing energy of 19,000 cm^{-1} and that 1 kJ mol^{-1} = 83 cm^{-1}).

A This involves comparing $10Dq$ with the pairing energy (PE) to determine that the complex is low spin. The crystal field splitting diagram shows the electron configuration from which the value of the CFSE can be calculated:

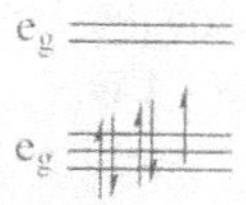

e_g

e_g

Strong-field case; low-spin d^5:

CFSE = $5(-2/5)\Delta_O + 2PE = -2(20,000/83) + 2(19,000/83) = -24.1$ kJ mol^{-1}

Remember to take into account the number of *additional* pairing energies required in low-spin cases before converting units from cm^{-1} to kJ mol^{-1} to give CFSE = −24.1 kJ mol^{-1}.

6.2.2 The Jahn–Teller Effect: Structural Distortions of Octahedral Complexes

According to the **Jahn–Teller theorem** a non-linear molecule in an electronically degenerate state will change its geometry to lower its symmetry, removing the degeneracy, and so attaining a lower energy. Molecules which do not have either one electron in each of a set of degenerate orbitals, or two electrons in each, are electronically degenerate. In the e_g case, the configurations e_g^1 and e_g^3 are electronically degenerate.

In complexes of general formula $[ML_6]$, certain d-electron configurations are associated with distortions from a regular octahedral structure. A particular example is found in complexes of the d^9 Cu^{2+} ion, which tend to have two *trans* bonds longer than the other four (Figure 6.6). This is an example of the Jahn–Teller effect. To understand how this works, it is necessary to consider the energy consequences of lowering the symmetry of an octahedral d^9 complex by moving two *trans* ligands on the z axis further from the metal, as shown in Figure 6.7. The reduced electronic repulsion along z results in lower energies for those orbitals with a z component, with a corresponding increase in energy for those orbitals without, to maintain the barycentre of each set. In the case of a d^9 electron configuration this results in an increase in the magnitude of the CFSE of $(\frac{1}{2})\delta_1$, so that the tetragonal structure is energetically more favourable, as found in $[Cu(NO_2)_6]^{4-}$ (Figure 6.6). This situation would not arise for a d^8 complex in which the two e_g orbitals are equally occupied, since the removal of the degeneracy would offer no energy benefit, one electron increasing in energy by $(\frac{1}{2})\delta_1$, the other decreasing in energy by the same amount. However, a low-spin d^7 complex should show this effect, although examples of this situation are rare. An octahedral high-spin complex of a d^4 metal ion such as Cr^{2+} or Mn^{3+} would have the electron configuration $t_{2g}^3e_g^1$, so should be subject to the Jahn–Teller effect, and spectroscopic measurements have identified examples in which this occurs. It might be expected that a metal ion with unequally occupied t_{2g} orbitals would also exhibit Jahn–Teller effects, low-spin d^4 metal ions for example. However, the t_{2g} orbitals interact less strongly with the ligands than the e_g orbitals, so the consequences are more difficult to observe.

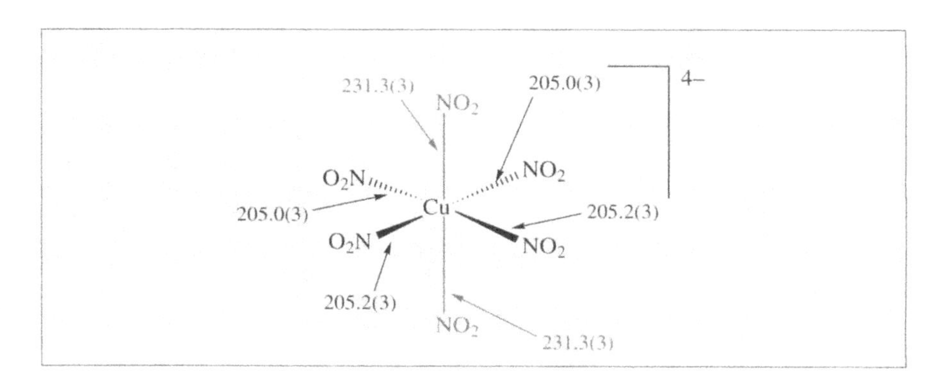

Figure 6.6 The structure of $[Cu(NO_2)_6]^{4-}$ with Cu–N bond distances shown in pm and standard deviations in parentheses

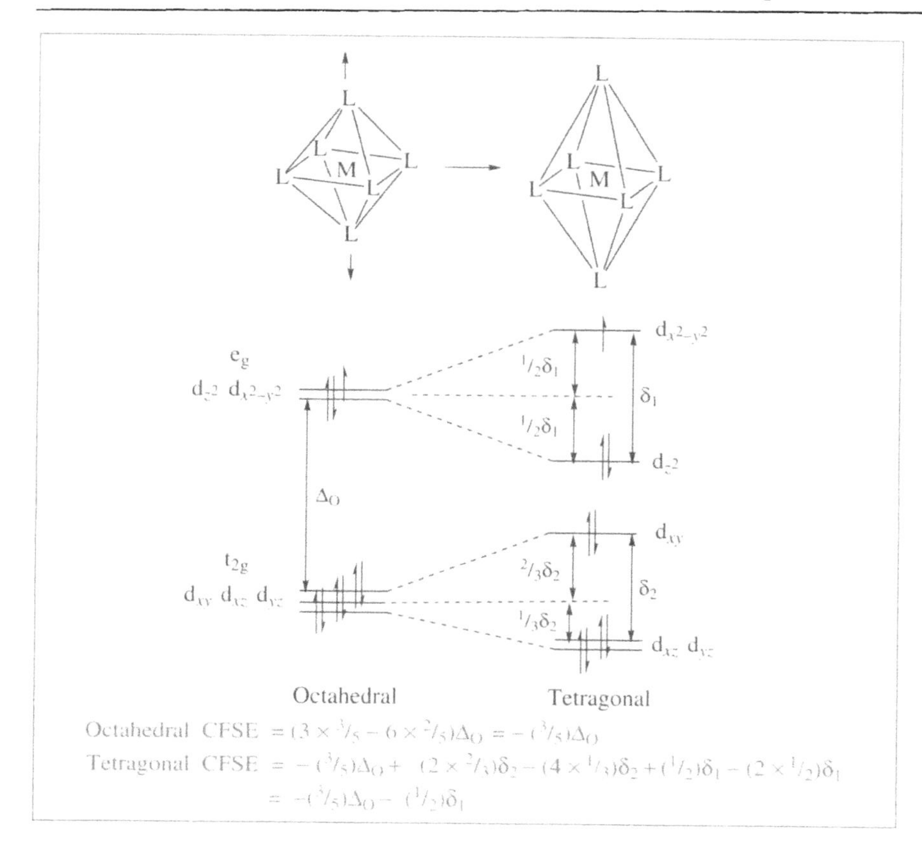

Figure 6.7 The Jahn–Teller effect and the effect of a tetragonal distortion of an octahedral crystal field on the energies of d orbitals

6.2.3 Tetrahedral Complexes

The procedure used to determine the crystal field splitting of the d orbitals in an octahedral complex may be applied similarly to tetrahedral complexes. In this case the process is perhaps simplest to understand if the ligand electron density is first localized from a spherical shell to the corners of a cube, as shown in Figure 6.8. In the set of coordinates used, the d_{z^2} and $d_{x^2-y^2}$ orbitals are oriented towards the centres of the faces of the cube, while the d_{xy}, d_{xz} and d_{yz} orbitals are oriented towards the centres of the edges. Geometrically it can be shown that, if the length of a cube edge is $2a$, then the distance from a cube corner to the centre of a face will be $\sqrt{2}a$. However, a corner will only be at a distance a from the centre of an edge. It might be expected from this simple geometric consideration that the repulsive force between the ligand electron density and electrons in the d_{z^2} and $d_{x^2-y^2}$ orbitals will be smaller than for the d_{xy}, d_{xz} and d_{yz} orbitals. Thus in the cubic structure the d_{z^2} and $d_{x^2-y^2}$ orbitals will move to lower energy, and the d_{xy}, d_{xz} and d_{yz} orbitals to higher energies. Once again the barycentre of the system must remain unchanged, so the d_{xy}, d_{xz} and d_{yz} orbitals will rise in energy by ⅖ of the splitting and the d_{z^2} and $d_{x^2-y^2}$ orbitals will drop in energy by

⅓ of the splitting. In order to derive a tetrahedral field from the cubic arrangement it is simply necessary to alternately remove four regions of electron density. The resulting splitting of the sets of d_{z^2} and $d_{x^2-y^2}$ and the d_{xy}, d_{xz} and d_{yz} orbitals is known as Δ_T and, for the same ligand L, calculations show that the magnitude of Δ_T for tetrahedral $[ML_4]$ is (4/9) of Δ_O for octahedral $[ML_6]$. In the tetrahedral situation, the symmetry label e is given to the d_{z^2} and $d_{x^2-y^2}$ orbitals and the symmetry label t_2 to the d_{xy}, d_{xz} and d_{yz} orbitals. Because the magnitude of Δ_T is only about half that of Δ_O, tetrahedral coordination compounds are essentially all high spin, although a low-spin tetrahedral organometallic complex of Co^{4+} is known: $[Co(nb)_4]$ (nb = norbornyl).

Tetrahedral symmetry labels: the tetrahedron does not have a centre of symmetry so the subscript labels $_g$ and $_u$ do not apply. The symmetry label e denotes the doubly degenerate set of orbitals d_{z^2} and $d_{x^2-y^2}$ and t_2 the triply degenerate set of orbitals d_{xy}, d_{xz} and d_{yz}.

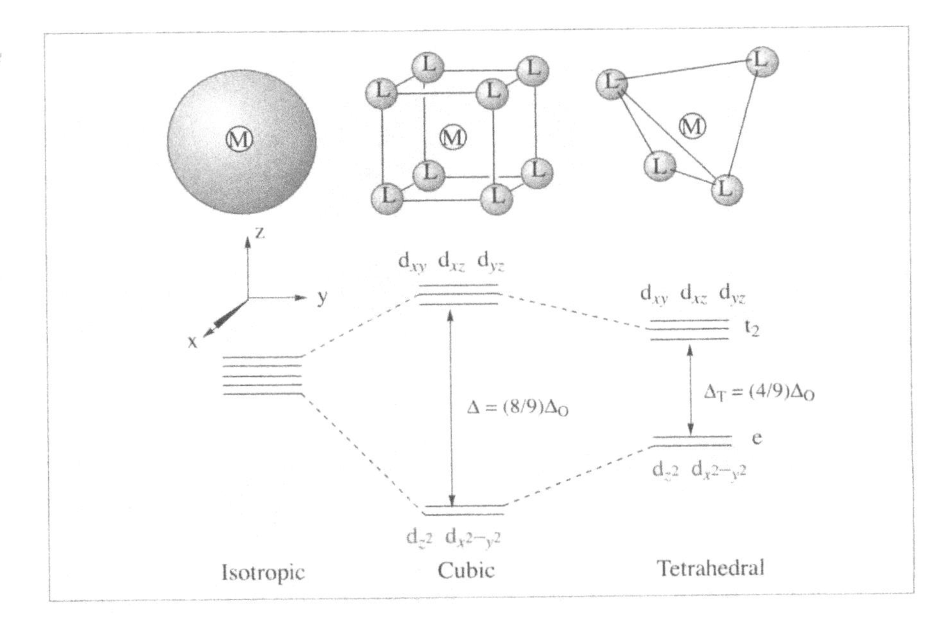

Figure 6.8 The effect of cubic and tetrahedral crystal fields on the energies of d orbitals

6.2.4 Square Planar Complexes

The energy level diagram for a square planar complex can be derived from that for an octahedral complex by considering the consequences of removing, to infinity, the two ligands bonded along the *z*-axis. In effect this is an extension to the limit of the tetragonal distortion described in Section 6.2.2, and again results in a lowering in the energy of those orbitals with a *z* component and a corresponding increase in the energies of those with no *z* component. In the square planar structure the d_{z^2} orbital may fall below the d_{xy} orbital in energy, so that the crystal field splitting Δ corresponds with the energy difference between the $d_{x^2-y^2}$ and d_{xy} orbitals as shown in Figure 6.9. Square planar complexes are most commonly associated with d^8 metal ions bound to strong-field ligands.

The reason for this can be seen from the energy level diagrams in Figure 6.9. In the presence of a weak-field ligand, a square planar structure offers no energy advantage over an octahedral structure. The splitting Δ is insufficient to induce pairing of the electrons in $d_{x^2-y^2}$ and d_{z^2} orbitals, so the benefit of moving the singly occupied d_{z^2} orbital to lower energy is cancelled by the penalty of moving the singly occupied $d_{x^2-y^2}$ orbital to higher energy. As in the tetragonal case, changes in the energies of the fully occupied t_{2g} orbitals also cancel. However, in the presence of a strong-field ligand where the energy difference between $d_{x^2-y^2}$ and d_{xy} exceeds the pairing energy (PE), it becomes energetically favourable to form a square planar complex, as the electrons become paired in d_{xy} and d_{z^2}, leaving the high-energy orbital $d_{x^2-y^2}$ unoccupied. An example of this effect is provided by d^8 Ni^{2+} which, with the weak-field ligand water, forms octahedral $[Ni(H_2O)_6]^{2+}$ but with the strong-field ligand CN^- forms square planar $[Ni(CN)_4]^{2-}$. Other metal ions which commonly form square planar complexes are Rh^+, Ir^+, Pd^{2+}, Pt^{2+} and Au^{3+}.

Square planar symmetry labels: the degeneracies of the e_g and t_{2g} orbital sets of an octahedral complex are lifted as the symmetry is reduced to square planar. The e_g orbitals give rise to non-degenerate orbitals of a_{1g} and b_{1g} symmetries, while the t_{2g} orbitals give rise to a non-degenerate orbital of b_{2g} symmetry and a doubly degenerate pair of orbitals of e_g symmetry, as shown in Figure 6.9. Those unfamiliar with group theory need not concern themselves with the origin of these labels, but should focus on which d orbitals they represent and their location in the energy level diagram.

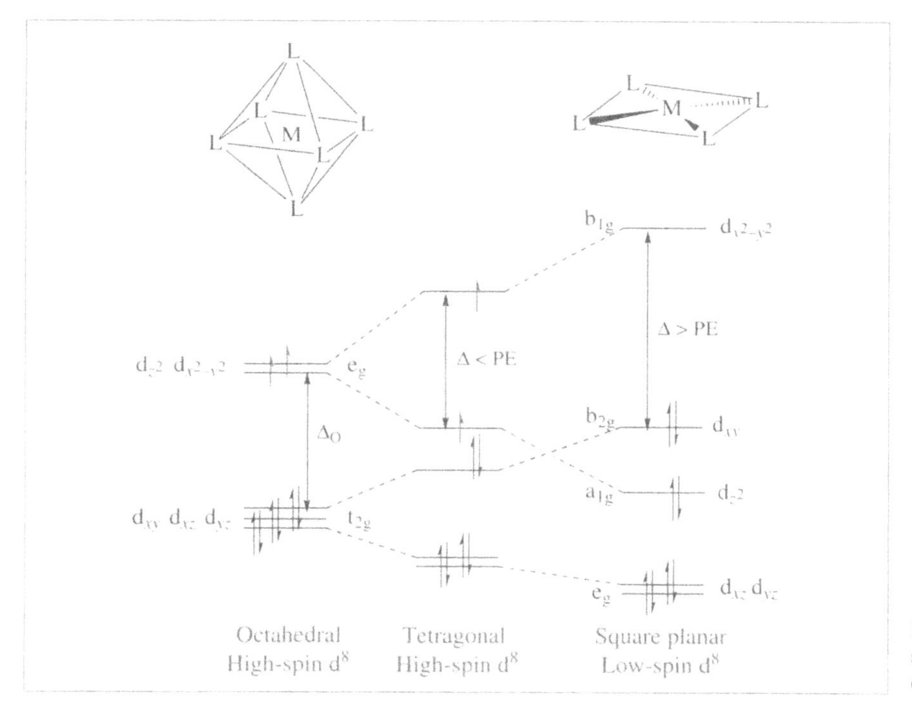

Figure 6.9 The effect of a square planar crystal field on the energies of d orbitals

6.2.5 Factors Affecting the Magnitude of the Crystal Field Splitting

The magnitude of the crystal field splitting experienced by a metal ion depends upon several factors (Box 6.3). Most obviously, the coordination geometry has an important effect, and we have seen that tetrahedral fields produce smaller splittings than octahedral fields involving the same ligands. The nature of the ligands is important in determining the exact value of the crystal field splitting, and so whether the complex is high or low spin. The nature of the metal centre is also important. Higher oxidation states with higher cationic charges tend to interact with ligands more strongly, leading to larger crystal field splittings. Furthermore, crystal field splittings tend to increase down groups. This is a result of the greater extension of the d orbitals in the second and third row of the d-block, which leads to a large interaction between the metal ion and the ligands. Thus, octahedral first-row d-block metal complexes show typical Δ_O values in the range 10,000–30,000 cm^{-1}, while for the second-row, values in the range 20,000–40,000 cm^{-1} are typical. In the f-block metal ions, the f orbitals are more core-like and crystal field effects are much smaller. Crystal field splittings of 50–100 cm^{-1} are more typical in the lanthanide series. Consequently, crystal field effects are of little importance in the chemistry of the lanthanide elements. The early actinides show somewhat larger crystal field splittings, with values in the range 1000–2000 cm^{-1}, although these are still small compared to d-block ions.

Box 6.3 Some Factors Affecting the Magnitude of Crystal Field Splitting

(a) *The coordination geometry and number of ligands.* For example, for VCl_4, Δ_T = 7900 cm^{-1}, and for $[VCl_6]^{2-}$, Δ_O = 15,400 cm^{-1} {in theory $\Delta_T = (4/9)\Delta_O$}.

Octahedral. Generally an octahedral coordination geometry might be expected to be the most favourable for a first-row d-block metal ion. However, a number of well-established exceptions exist.

Tetrahedral. Where there is little or no CFSE difference between octahedral and tetrahedral geometries, the tetrahedral structure may be favoured owing to the lower charge and ligand–ligand repulsions associated with four rather than six coordination, *e.g.* $[MnX_4]^{2-}$ (X^- = halide) contain weak-field ligands with d^5 configurations, giving CFSE = 0. Zinc(+2) complexes are typically tetrahedral due to the smaller ionic radius and d^{10} configuration of Zn^{2+}. Cobalt(+2) also forms tetrahedral complexes with weak-field ligands owing to the small CFSE difference between octahedral and tetrahedral geometries.

Tetragonal. This geometry normally arises from Jahn–Teller distortion of an octahedral structure, and is best illustrated by complexes of Cu^{2+} which, if octahedral, would involve unequally occupied e_g levels owing to the d^9 configuration. In such cases a tetragonal structure is energetically more favourable. Tetragonal structures are most readily observed in six-coordinate complexes of d^9 and high-spin d^4 ions (see Section 6.2.2).

Square planar. This geometry normally arises with d^8 metal ions such as Ni^{2+} or Rh^+ when bound to strong-field ligands. In such cases the square planar geometry is favoured energetically over octahedral. The reasons for this appear similar to those underlying the formation of tetragonal complexes of d^9 ions through Jahn–Teller distortion. However, the d^8 configuration is not electronically degenerate in octahedral symmetry, so the Jahn–Teller theorem does not apply in this case (see Section 6.2.4).

(b) *The nature of the ligands*: strong, intermediate or weak field. The approximate order of ligand field strengths is given by the spectrochemical series (see Section 6.2.6), *e.g.* for $[Ru(H_2O)_6]^{2+}$, $\Delta_O = 19,800$ cm^{-1}, but for $[Ru(CN)_6]^{4-}$, $\Delta_O = 33,800$ cm^{-1}; also for $[Rh(H_2O)_6]^{3+}$, $\Delta_O = 27,200$ cm^{-1}, but for $[Rh(NH_3)_6]^{3+}$, $\Delta_O = 34,100$ cm^{-1}.

(c) *The oxidation state of the metal*: higher oxidation states lead to larger splitting, *e.g.* for $[Co(H_2O)_6]^{2+}$, $\Delta_O = 9200$ cm^{-1}, but for $[Co(H_2O)_6]^{3+}$, $\Delta_O = 20,760$ cm^{-1}; also for $[Ru(H_2O)_6]^{2+}$, $\Delta_O = 19,800$ cm^{-1}, but for $[Ru(H_2O)_6]^{3+}$, $\Delta_O = 28,600$ cm^{-1}.

(d) *The nature of the metal centre*: splittings tend to increase down groups. Metal ions may be placed in an approximate order, according to increasing crystal field splitting for a given ligand and complex type: $Mn^{2+} < Ni^{2+} < Co^{2+} < Fe^{2+} < V^{2+} < Fe^{3+} < Cr^{3+} < V^{3+} < Co^{3+} < Mn^{4+} < Rh^{3+} < Pd^{4+} < Ir^{3+} < Pt^{4+}$. Thus for $[Fe(H_2O)_6]^{3+}$, $\Delta_O = 14,000$ cm^{-1}, but for $[Ru(H_2O)_6]^{3+}$, $\Delta_O = 28,600$ cm^{-1}; and for $[Co(H_2O)_6]^{3+}$, $\Delta_O = 20,760$ cm^{-1}, but for $[Rh(H_2O)_6]^{3+}$, $\Delta_O = 27,200$ cm^{-1}

Worked Problem 6.3

Q Explain why $[Mn(H_2O)_6]^{2+}$ reacts with CN^- to form $[Mn(CN)_6]^{4-}$ which has one unpaired electron, but with I^- to give $[MnI_4]^{2-}$ which has five unpaired electrons.

A Since CN^- is a strong-field ligand, with d^5 Mn^{2+} it gives a low-spin octahedral complex with a $t_{2g}^5e_g^0$ electron configuration having one unpaired electron. Since I^- is a weak-field ligand, with d^5 Mn^{2+} an octahedral complex, $[MnI_6]^{4-}$, would have a $t_{2g}^3e_g^2$ configuration with zero CFSE. Thus there is no CFSE penalty in going to the tetrahedral structure of $[MnI_4]^{2-}$. However, there is the benefit of a lower anionic charge and less steric interaction between the large iodide ligands, so tetrahedral $[MnI_4]^{2-}$ is formed in preference to $[MnI_6]^{4-}$. Tetrahedral complexes are high spin, so five unpaired electrons are expected.

6.2.6 Limitations of the Crystal Field Model

Although the CFT model is successful in explaining many of the properties of transition metal complexes, it fails to account for some important observations. In particular, spectroscopic measurements have been used to define a relative order for ligands in terms of the magnitude of the crystal field splitting they induce. This order is known as the spectrochemical series:

Weak field

$$I^- < Br^- < S^{2-} < SCN^- < Cl^- < N_3^-, F^- < (NH_2)_2C{=}O, OH^- < O_2CCO_2^{2-} < O^{2-} < H_2O < NCS^- < NC_5H_5\ (py) < NH_3 < 2,2'\text{-}NC_5H_5C_5H_5N\ (bpy) < NO_2^- < CH_3^- < CN^- < CO$$

Strong field

This order is based on measurements of complexes with a variety of metal ions and so represents an averaged ordering which may not be exactly correct for any given metal ion but represents the general picture. Bearing in mind that the crystal field model is an ionic model based on the electrostatic effect of the ligands on the energies of the metal ion d orbitals, it is surprising to find that anionic ligands such as Br^-, F^- and OH^- should be weaker-field than neutral ligands such as NH_3, PR_3 and CO. This anomaly arises because the simple crystal field model does not take account of covalency in metal–ligand bonding. Modifications to the CFT model in the LFT model take account of these effects, but use of an MO description of bonding naturally includes allowance for covalency and so resolves this problem in a more general sense.

6.3 The Molecular Orbital Model of Bonding

6.3.1 Octahedral Transition Metal Complexes

The electronegativities of transition metal ions show that it is unrealistic to regard the metal–ligand bond in transition metal complexes as purely ionic. MO theory naturally allows for a range of bonding types varying from largely ionic to largely covalent. As with CFT, symmetry is at the heart of the MO model, and group theory underlies the construction of an MO description of bonding. The approach taken here will not explicitly use the mechanism of group theory but the derivation of the symmetry labels of the various allowed combinations of ligand orbitals originate from this approach (Box 6.4).

Box 6.4 Energy Level Diagrams

In order to develop a qualitative MO energy level diagram to describe bonding between a transition metal ion and a set of attached ligands, it is first necessary to define the symmetry properties of the various sets of atomic orbitals which will be used to construct the MOs. Group theory provides a means of determining the symmetries of orbital combinations allowed by the structure of the molecule. These are identified by symmetry labels such as t_{2g} or e_g. Those unfamiliar with group theory should simply take the symmetry labels as given. To those familiar with group theory they will have a deeper significance.

In constructing MO energy level diagrams for complexes we will only consider the metal atom and the ligand donor atoms, X, directly bound to it (in a complex $[ML_n]$, X = L when L is monoatomic, *e.g.* F^-, otherwise X is the donor atom within L, *e.g.* N in NH_3). Each atom X will have an s orbital and three p orbitals in its valence shell (Figure 6.10) and it is convenient to regard one p orbital on each X as pointing directly at the metal in a radial fashion as if to form a σ bond. These radial p orbitals are given the label p_r, and as a group will will have the same symmetry properties as the group of s orbitals on X (Figure 6.10a). As a consequence, these orbitals can mix, in effect forming two sp hybrids, one directed towards the metal which may be used in s bonding and the other directed away from the metal. This second σ-type orbital may either be used in bonding to other ligand atoms, as in OH^- or NH_3, or will be occupied by a non-bonding lone pair of ligand electrons, as in F^-. In addition, there are two further p orbitals on X oriented perpendicular to p_r and each to the other (Figure 6.10b). These will be referred to as the tangential orbitals p_t, and may be used in π bonding interactions if their symmetries correspond with those of metal valence shell orbitals (Figure 6.10c). Some ligands, such as CO, contain empty π* orbitals which have the same symmetry properties as the p_t orbitals, but lie at higher energy. These may also interact with filled metal valence shell orbitals (Figure 6.10d and e).

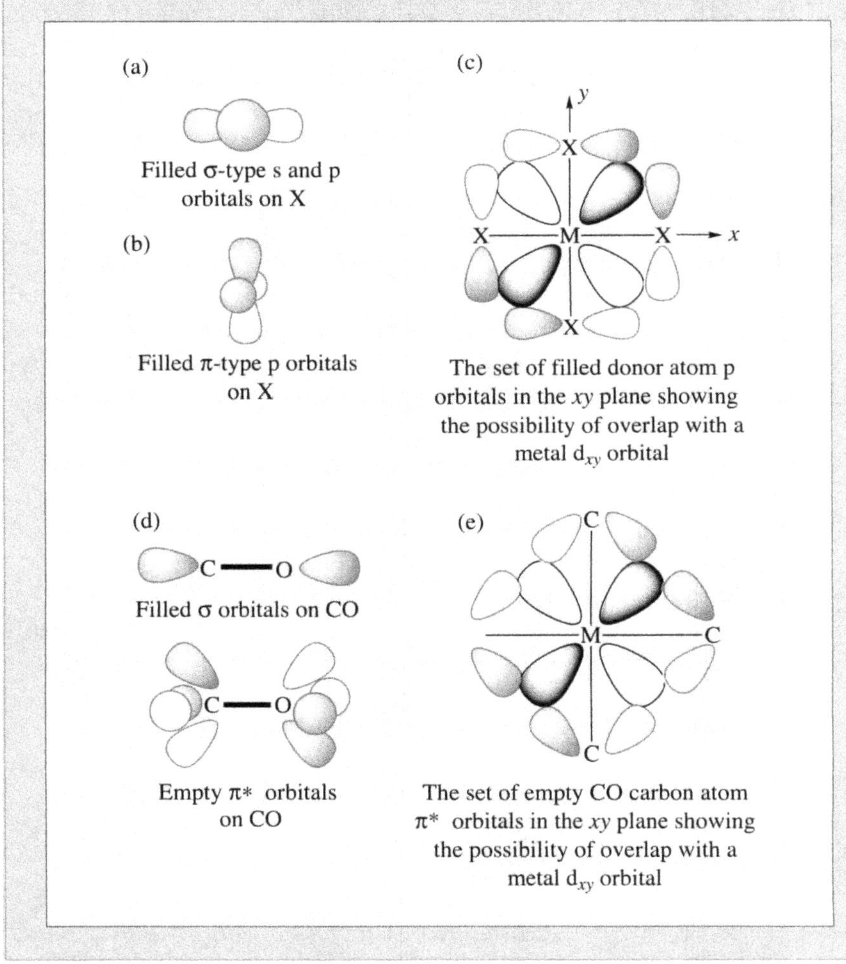

Figure 6.10 (a) The atomic orbitals of a donor atom X in a ligand L in an octahedral complex. (b) The tangential atomic orbitals p_t of a donor atom X. (c) A π interaction between the p_t orbitals of X and a d orbital of m in $[ML_6]$. (d) σ and π* orbitals in CO. (e) π interactions between metal d_{xy} and CO π* orbitals.

In the MO bonding model, atomic orbitals which are of the same symmetry, similar energy and which have significant overlap are combined to produce an equal number of MOs. These are no longer localized at a particular atom but may be distributed over several atoms. In the simplest case of the hydrogen molecule, the two hydrogen 1s atomic orbitals are combined using the linear combination of atomic orbitals (LCAO) approach to produce two molecular orbitals; one bonding MO of lower energy and one antibonding MO of higher energy. In the case of an octahedral complex of a transition metal ion the situation is more complicated, because there are nine metal valence shell orbitals, one s, three p and five d, as well as six ligand σ-type orbitals to consider (Figure 6.10). However, the basic approach is the same and involves the interaction of combinations of ligand orbitals with metal valence shell orbitals of the same symmetry to form bonding or antibonding MOs.

Firstly it is necessary to construct, from the six σ-type ligand donor atomic orbitals, six new combination orbitals known as **ligand group orbitals** (LGOs). These LGOs must have symmetries allowed by the structure of the molecule. They can be constructed from the six ligand σ-type orbitals using the LCAO approach, and a knowledge of the allowed symmetries obtained from group theory. In the octahedral case the allowed LGOs have the symmetry labels a_{1g}, e_g and t_{1u}, and are shown in Figure 6.11. These six LGOs have appropriate symmetries to mix with six of the metal valence shell orbitals, which also have a_{1g}, e_g and t_{1u} symmetries, to form σ bonding and antibonding combinations. These are shown in Figures 6.12, 6.13 and 6.14 for each interaction respectively. The remaining three metal atomic orbitals, d_{xy}, d_{xz} and d_{yz}, are not of the correct symmetry to interact with the σ LGOs and contribute a triply degenerate set of non-bonding MOs having the symmetry label t_{2g}. The result of incorporating the 15 MOs created from the nine valence shell and six ligand σ-type orbitals in a single MO energy level diagram is shown in Figure 6.15. Although the relative energies of the orbitals cannot be known with certainty from this qualitative approach, several assumptions may reasonably be made. Firstly, the metal valence shell orbitals will be higher in energy than the filled σ-type orbitals of the more electronegative ligand donor atoms. Secondly, the MOs can be arranged in sets with the energy order bonding orbitals below non-bonding orbitals below antibonding orbitals. Finally, it might be expected that the metal valence shell *n*s and *n*p orbitals will extend further than the $(n - 1)$d orbitals, and so show greater overlap and stronger interaction with the LGOs.

Hybridization in an octahedral complex: in the valence bond model of coordination compounds the formation of six bonds to an octahedral transition metal ion would involve six d^2sp^3 hybrid orbitals. In the approach used here the differing symmetry properties of the s, the p_x, p_y and p_z and the $d_{x^2-y^2}$ and d_{z^2} orbitals used in bonding are explicitly taken into account. However, the s, all of the p and two of the d metal valence shell orbitals are still used in σ-bond formation.

The MO diagram shown in Figure 6.15 for a d^3 ion contains six σ bonding orbitals, a_{1g}, e_g and t_{1u}, which have a capacity of 12 electrons. This matches the number of electrons donated by six Lewis base ligands in $[ML_6]$. Above these in energy lies the non-bonding t_{2g} orbitals which, in this σ bonding model are entirely metal in character. Next there are two antibonding e_g* orbitals which, with the t_{2g} orbitals set comprise the MO equivalent of the e_g and t_{2g} levels in the CFT model. Thus the MO

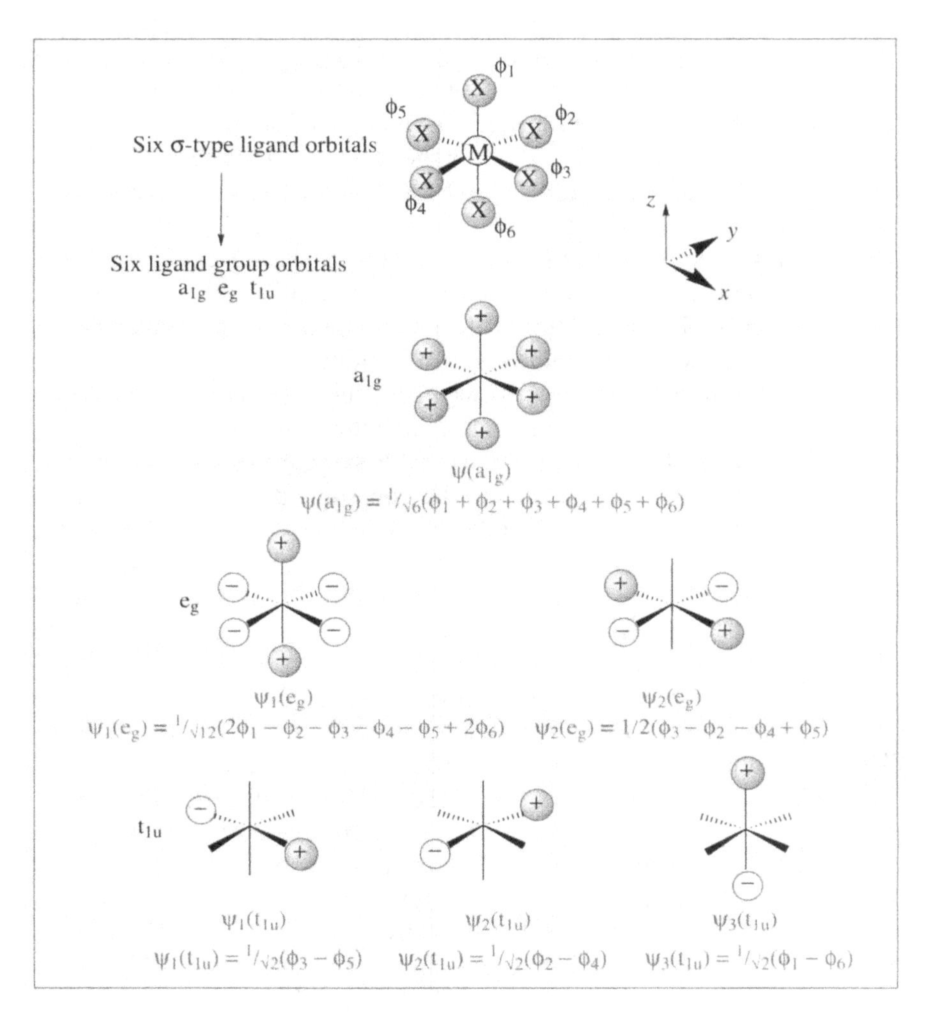

Figure 6.11 The σ ligand group orbitals (LGOs) for an octahedral complex, $[ML_6]$

model contains the CFT model, but there is an important difference. In the MO model, occupation of the antibonding e_g* orbital is seen as opposing bonding arising from occupancy of the lower energy bonding e_g level. Thus an octahedral complex of a d^{10} ion such as Zn^{2+}, in which e_g* is fully occupied by four electrons, would have an effective σ-bond order of four not six. In practice, complexes of d^{10} ions are normally tetrahedral, not octahedral. This corresponds with the use of only the s and p valence shell orbitals in σ bonding to the ligands, the d-orbitals in effect constituting a filled core subshell.

Although the MO model developed so far contains the features found in the CFT model, it does not yet account for the finding that anionic F^- is a weak-field ligand but neutral CO is strong field. To do this it is necessary to include in the model other orbitals associated with the ligand donor atoms, particularly those which can participate in π bonding.

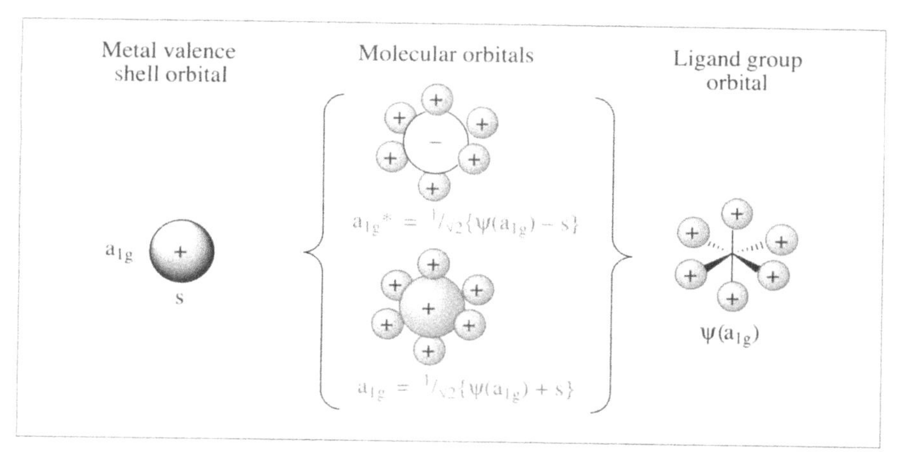

Figure 6.12 The a_{1g} M–L σ bonding and σ* anti-bonding MOs in an octahedral complex

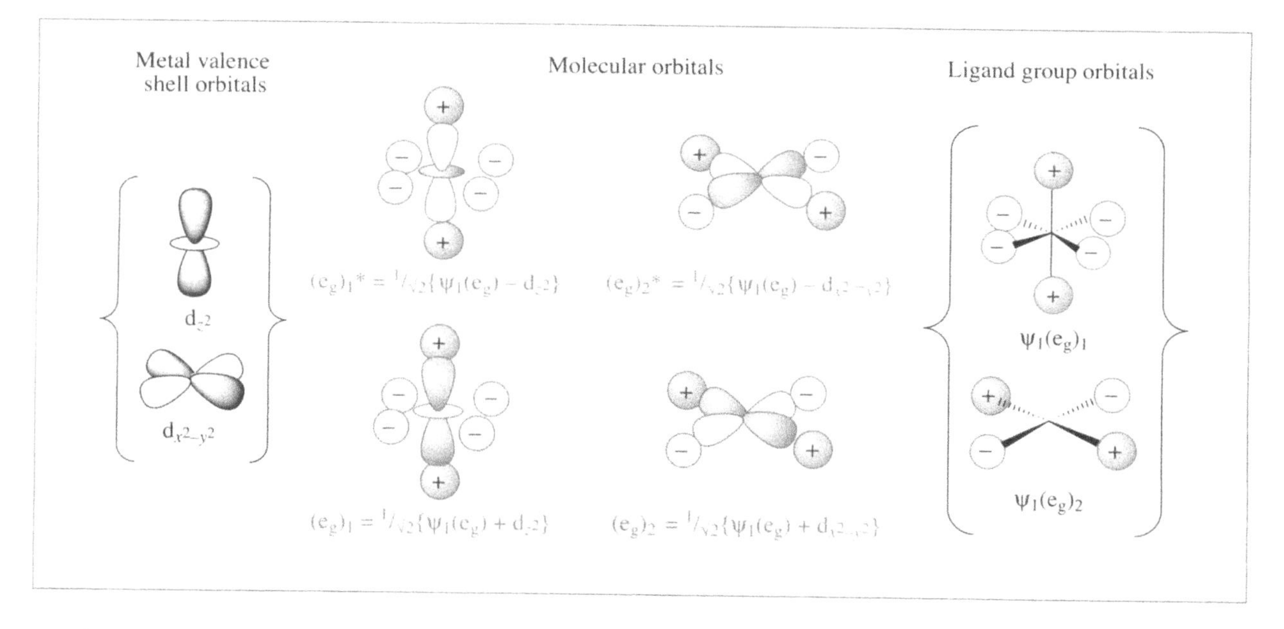

Figure 6.13 The e_g M–L σ bonding and anti-bonding MOs in an octahedral complex

Twelve LGOs can be constructed from the 12 donor atom π-type p_t orbitals in an octahedral complex. Out of these t_{1g}, t_{2g}, t_{1u} and t_{2u} symmetry LGOs, only the t_{2g} set of three is of suitable symmetry to interact with any of the metal d orbitals. This interaction produces t_{2g} and t_{2g}^* MOs, as shown in the fragment MO energy level diagram in Figure 6.16a, which, for the sake of simplicity, only presents a part of the complete MO energy level diagram. In the case of a ligand such as F^-, which has low-energy filled p atomic orbitals, the lower energy t_{2g} MO has more ligand than metal character and is fully occupied because of the six ligand p electrons donated into the MO scheme. Thus the 'metal electrons' must be accomodated in the t_{2g}^* orbital which is more metal than ligand in character. This t_{2g}^* MO is at a higher energy than its counterpart

Metal and ligand electrons
since electrons are indistinguishable, it is meaningless to refer to 'metal electrons' as if we could know which come from the metal and which from the ligand. However, since the six electrons in the t_{2g} LGOs must be accommodated in the MO scheme, as well as those originating from the metal, it is convenient to think of the 'ligand electrons' as occupying the t_{2g} and the 'metal electrons' occupying the t_{2g}^* orbitals.

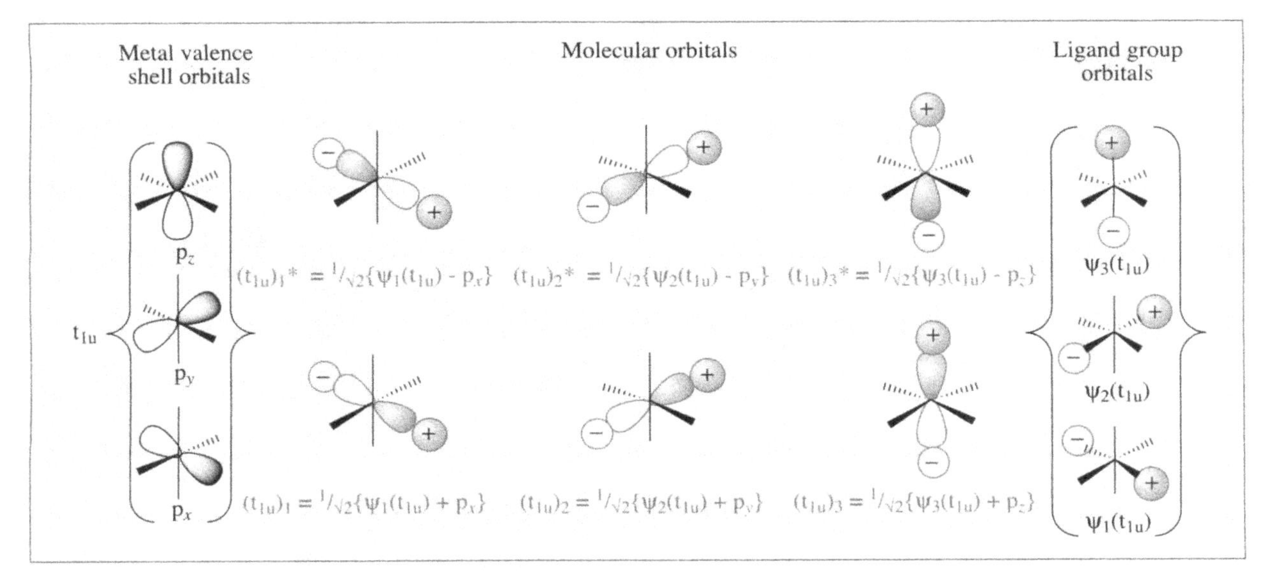

Figure 6.14 The t_{1u} symmetry M–L σ bonding and anti-bonding MOs in an octahedral complex

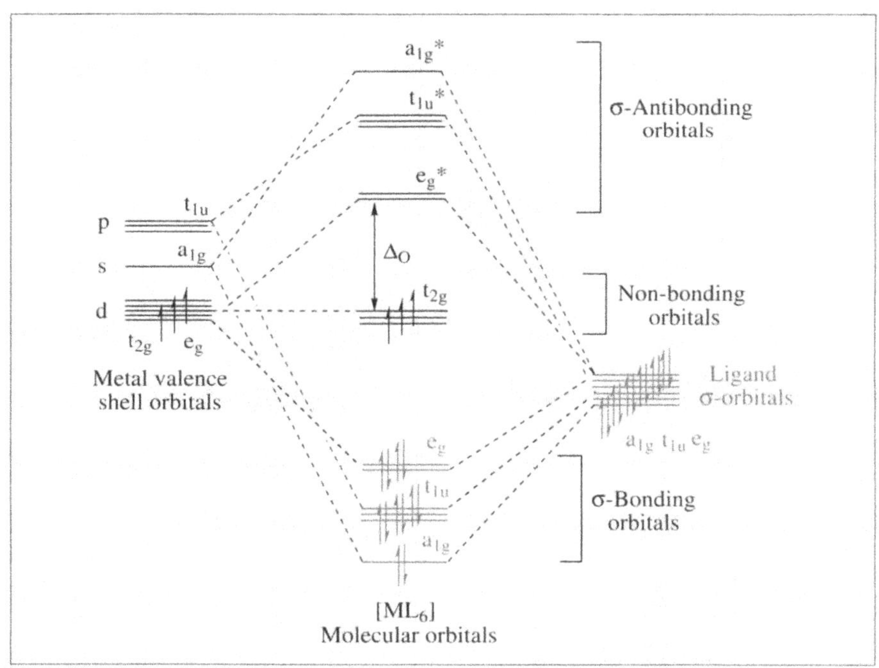

Figure 6.15 The MO energy level diagram for σ bonding in an octahedral complex, $[ML_6]$

Because the F^- ion is donating charge into the MO scheme π orbitals it is refered to as a π-donor ligand. However, since the bonding interaction has some ionic character, it is fair to say that the t_{2g} orbital has more ligand than metal character and that the t_{2g}^* orbital more metal than ligand character.

in the σ bonding model, so the magnitude of Δ_O is reduced when π bonding with a π donor ligand such as F^- is taken into account. In this way the MO model can now explain the observation that F^- is a weak-field ligand.

In the case of a ligand like CO the π-type p_t orbitals on carbon are involved in π bonding to oxygen. However, there is a set of 12 unoccu-

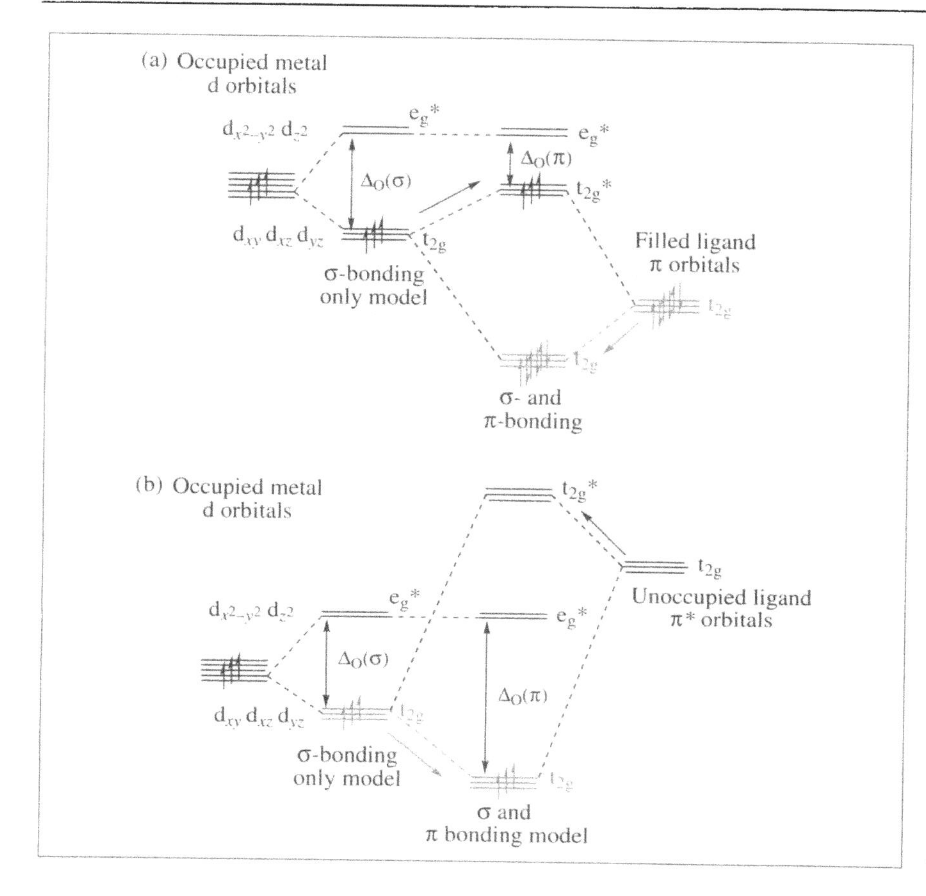

Figure 6.16 Fragment MO energy level diagrams for π bonding in an octahedral complex, $[ML_6]$: (a) the effect of π donor ligands; (b) the effect of π acceptor ligands

pied π* MOs on CO which are of the same symmetries as the set of donor atom p orbitals (Figure 6.10d and e), so once again a set of 12 LGOs can be obtained from the π* orbitals. As before, the three t_{2g} LGOs can interact with the metal t_{2g} orbitals to produce t_{2g} and t_{2g}* MOs (Figure 6.16b). However, in this case the lower energy t_{2g} MO will be more metal in character and the higher energy t_{2g}* orbital will be more ligand in character. As the proligand π* orbitals were unoccupied, the 'metal electrons' can now occupy the lower energy t_{2g} orbital, so that the magnitude of Δ_O is increased when π bonding with a π acceptor ligand such as CO is taken into account. In this way the MO model can explain the observation that, although neutral, CO is a strong-field ligand. More complete MO diagrams can be drawn which include the other non-bonding ligand π-type LGOs as shown in Figure 6.17, and these are useful in interpreting the electronic spectra of octahedral complexes which are considered in the next chapter.

A further point which emerges from this consideration of π bonding is that, in the presence of π donor ligands such as F^-, the 'metal electrons' move to higher energy in the t_{2g}* orbital and so are easier to ionize. The

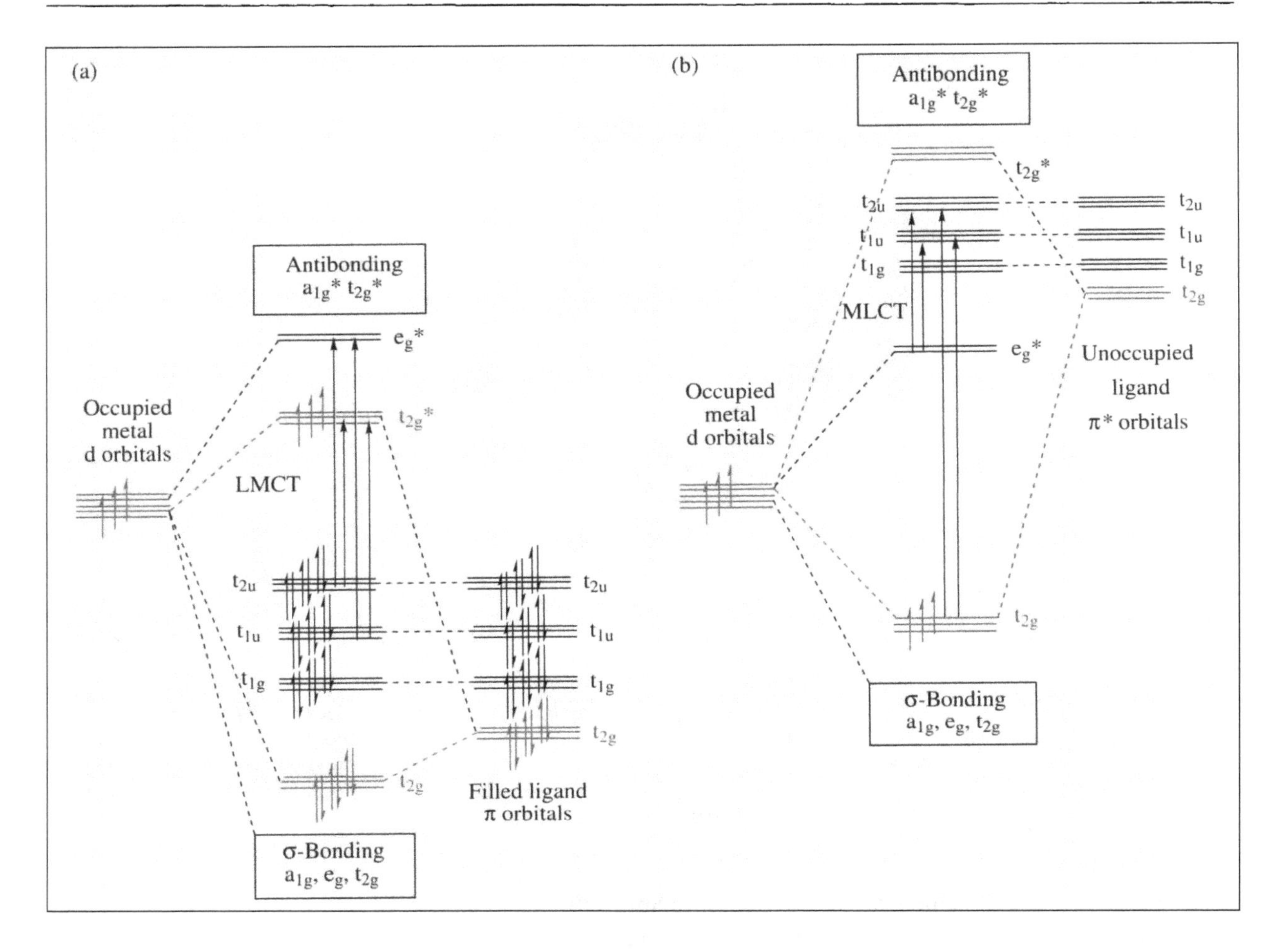

Figure 6.17 MO energy level diagrams which include non-bonding ligand π orbitals for an octahedral complex, $[ML_6]$: (a) π donor ligands; (b) π acceptor ligands. The terms LMCT and MLCT are discussed in Section 7.2.2

result is that π donor ligands tend to stabilize higher oxidation states. As an example, the Co^{4+} complex $[CoF_6]^{2-}$ can be isolated even though Co(+4) is an unusually high oxidation state for cobalt. Conversely, in complexes of ligands with empty π-type orbitals the 'metal electrons' move to lower energy in the t_{2g} MO so that lower oxidation states are stabilized. Thus, although Cr^{2+} compounds in general are rapidly oxidized by air, the lower oxidation state Cr(0) complex $[Cr(CO)_6]$ containing the π acceptor ligand CO is relatively stable to oxidation in air. Following the Lewis model, π acceptor ligands are sometimes referred to as π-acids and π donor ligands as π-bases.

6.3.2 Synergic Bonding and the Electroneutrality Principle

In the MO model we have considered metal–ligand bonding in octahedral complexes in order to see the effects of π bonding on the crystal field splitting parameter. However, it is sometimes simpler to look at a single metal–ligand unit when considering metal–ligand interactions. This is helpful in discussing the electroneutrality principle which requires that, after taking into account electron donation from the ligands, the central metal ion in a complex should be more or less neutral in charge. In a purely covalent σ-bond between a metal ion and a ligand donor atom, the two electrons might be apportioned one to the metal and one to the ligand. This would imply that a Co^{3+} ion with six ligands, *e.g.* in $[Co(NH_3)_6]^{3+}$, would receive one electron from each ligand and attain a net charge of −3, violating the electroneutrality principle. However, dative bonds are normally quite polar and the ligand donor atoms, being the more electronegative, would retain more electron density. Thus the net transfer of half an electron per ligand to give a neutral cobalt centre, as required for electroneutrality, does not seem too unreasonable. Unfortunately, this argument does not work with low oxidation state complexes such as $[Cr(CO)_6]$. Here the metal centre is already neutral, being in oxidation state 0, so any donation of charge from the ligand would seem to be unacceptable. However, if the ligand is able to accept charge from occupied metal d orbitals into its unoccupied π* orbitals, the possibility exists that σ-donation from the ligand to the metal could be compensated, at least in part, by π back-donation from the metal to the ligand. This is shown schematically for a single metal–ligand interaction in Figure 6.18a. Such an interaction is known as synergic bonding, since the σ-donation of charge from the ligand is reinforced by π back-donation from the metal to the ligand. In a valence bond model this might be represented by contributions to bonding from $M^-–C≡O^+$ and M=C=O resonance forms. In the MO model it corresponds with the lower energy t_{2g} orbitals, in which the 'metal electrons' reside, having some ligand character so that charge appears to expand from the metal onto the ligands. This metal electron cloud expansion is known as the nephelauxetic effect. It is observed in spectroscopic measurements, which show that metal ion electron–electron repulsions in some complexes are significantly lower than in the free uncomplexed metal ion. The lowering in the energy of the occupied t_{2g} MOs through interaction with ligand π* orbitals constitutes a metal–ligand bonding interaction, even though the ligand π* orbitals are antibonding with respect to the C–O interaction. Thus back donation of charge from the metal to the ligand strengthens the metal–ligand bond, but weakens the C–O bond through the population of orbitals which are antibonding within C–O. Although CO has been used here as an example, many other ligands can enter into synergic bonding interactions with metal ions. Ligands which are isoelectronic with CO include N_2, NO^+, CN^- and $RC≡C^-$. Alkenes are also able to enter into synergic bonding interactions with metal ions. A simple example is provided by the ethene complex $[PtCl_2(H_2C=CH_2)]$. In

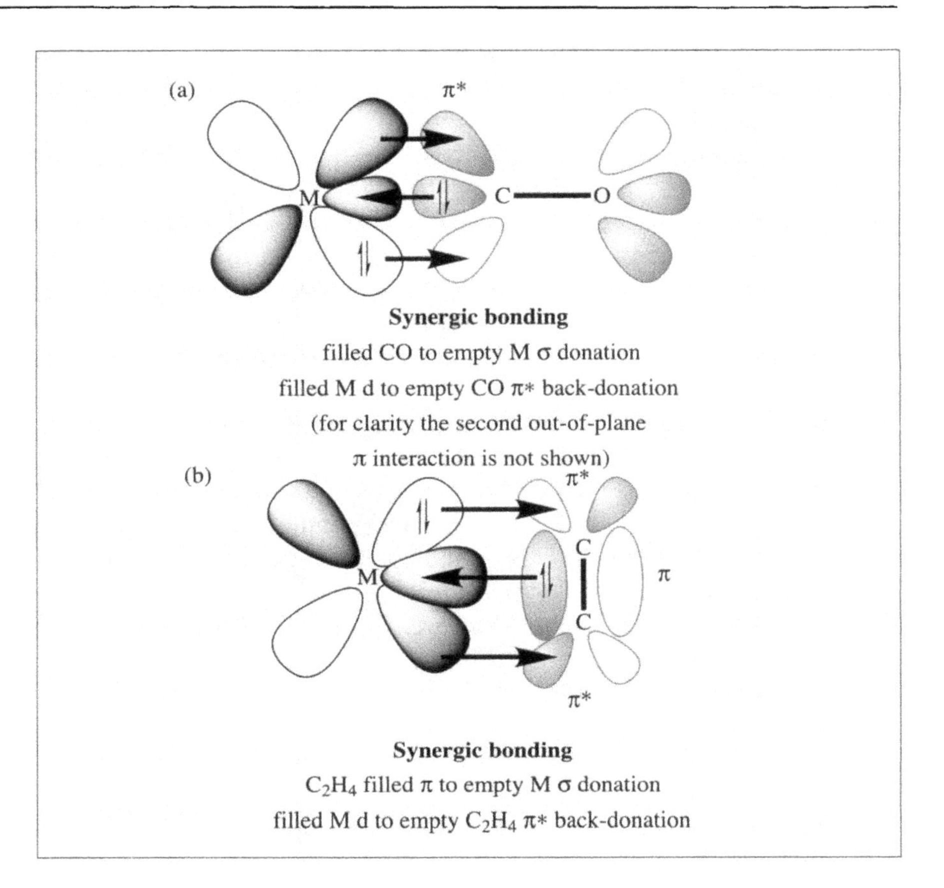

Figure 6.18 Examples of synergic bonding interactions: (a) in carbonyl complexes; (b) in alkene complexes

this complex the filled π orbital of ethene acts as a donor towards an empty σ-type metal orbital. This σ interaction is reinforced by π back-bonding from a filled metal π-type orbital to the ethene π* orbital, as shown in Figure 6.18b. Another group of ligands which show some back-bonding capacity are the organophosphines, R_3P. Early models of bonding by these ligands invoked empty d orbitals on phosphorus as the π acceptor orbitals, but more recent studies show that certain combinations of σ* orbitals are better placed to interact with filled π-type orbitals on the metal. Compared to CO, which is a weak σ donor ligand but a strong π acceptor, organophosphine ligands are strong σ donors but rather weak π acceptors.

The experimental evidence for the synergic bonding comes from several sources. Perhaps the most direct evidence comes from infrared spectroscopic studies of coordinated ligands, among which CO provides a good example. In free CO the C–O stretching frequency, $\nu_{max}(CO)$, is 2147 cm^{-1}, arising from a valence bond C–O bond order of three for the $^-C{\equiv}O^+$ bonding configuration $\sigma^2,\pi^4,\pi^{*0},\sigma^{*0}$. The value of $\nu_{max}(CO)$ is a measure of the strength of the C–O bond and, as this is reduced by back donation, so the value of $\nu_{max}(CO)$ should fall. This is exactly what is

Table 6.1 CO stretching frequencies in some binary metal carbonyls

Complex	Oxidation state	ν_{max}(CO) (cm^{-1})	CN
$[Mn(CO)_6]^+$	+1	2090	6
$[Cr(CO)_6]$	0	2000	6
$[V(CO)_6]^-$	−1	1860	6
$[Ni(CO)_4]$	0	2060	4
$[Co(CO)_4]^-$	−1	1890	4
$[Fe(CO)_4]^{2-}$	−2	1790	4
Free CO		2147	

found in binary metal carbonyl complexes, and the lower the formal oxidation state of the metal, the lower the value of ν_{max}(CO) (Table 6.1). The lower the metal oxidation state, the more it is able to donate charge into the π^* orbitals of the CO ligand and the lower the C–O bond order. Similar observations can be made for other ligands of this type, *e.g.* N_2, NO, CN^- and $RC{\equiv}C^-$. The other expectation from the synergic bonding model is that the metal–ligand bond order should increase. Evidence in support of this can be obtained from NMR spectroscopy studies of metal–organophosphine complexes (Figure 6.19). In these compounds the coupling constants, $^1J(M,^{31}P)$, between the ^{31}P and metal nuclei, depend on the electron density in the metal–phosphorus bond, more electron density being associated with a larger value for $^1J(M,^{31}P)$. In $[PtCl_2(PEt_3)_2]$, $^1J(^{195}Pt,^{31}P)$ is higher in the *cis* isomer, in which a π-donor chloride ligand is *trans* to the M–P bond, than in the *trans* isomer, where the two π-acceptor phosphine ligands are mutually *trans* and compete for metal electron density.

Bond distances obtained by single-crystal X-ray diffraction structural studies can provide another source of information about synergic bonding in transition metal complexes. However, such data need to be used with care, since changes in bond lengths are small and may be less than the experimental errors in determining the bond distances.

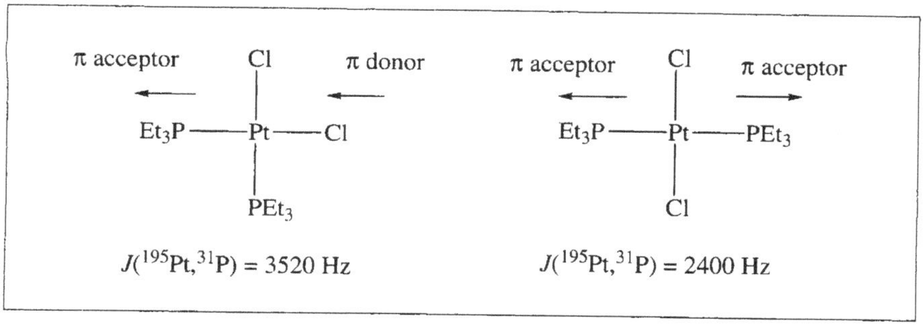

Figure 6.19 An example of the effect of π bonding on metal–phosphorus NMR coupling constants

Worked Problem 6.4

Q The ^{31}P NMR spectrum of *cis*-$[PtCl_2(PEt_3)_2]$ shows a coupling constant $J(^{195}Pt,^{31}P)$ of 3520 Hz, but in *trans*-$[PtCl_2(PEt_3)_2]$ a value of 2400 Hz is found for $J(^{195}Pt,^{31}P)$. Explain this observation.

A

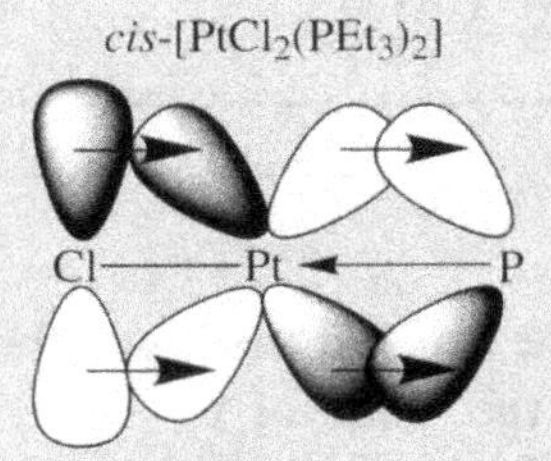

Pt–P bonding reinforced by π donation from Cl^-

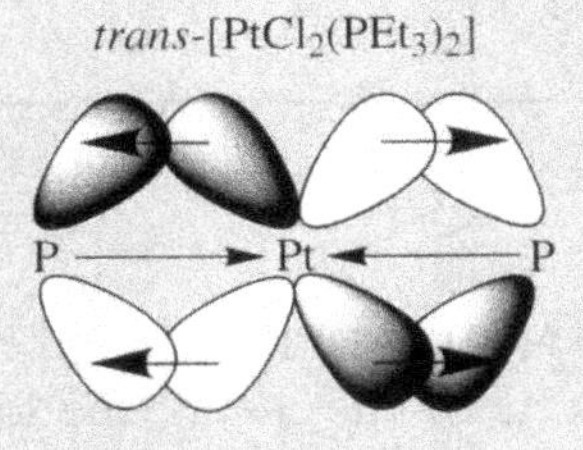

P atoms compete for electron density on Pt

Since Cl^- is a π donor ligand, this reinforces back bonding from a filled metal d orbital to an empty *trans* phosphorus d or σ* orbital. This results in higher electron density in the Pt–P bond, and a higher $^1J(^{195}Pt,^{31}P)$ coupling constant than where two *trans* phosphorus atoms compete for the charge density on Pt.

6.3.3 Tetrahedral Complexes

The MO energy level diagrams for tetrahedral and square planar complexes can be developed following a similar procedure to that used above for an octahedral complex. Again, the use of group theory allows the symmetries of the LGOs to be determined. In the case of a tetrahedral complex, the four σ bonding LGOs have a_1 and t_2 symmetries. These allow interactions with the metal valence shell s (a_1) and p_x, p_y and p_z (t_2) orbitals, as well as the metal d_{xy}, d_{xz} or d_{yz} (t_2) orbitals (Figure 6.20). The mixing of the three triply degenerate sets of t_2 orbitals results in three low-energy bonding MOs (t_2) associated with metal–ligand σ bonding, three MOs at intermediate energy (t_2') corresponding with the t_2 levels of the crystal field model, and three antibonding MOs (t_2*) at high energy. In this scheme the metal $d_{x^2-y^2}$ and d_{z^2} orbitals are of e symmetry and contribute a pair of non-bonding MOs as there are no ligand σ orbitals of this symmetry. Once again, the MO model contains the energy levels of the crystal field model lying above a set of four σ bonding orbitals which can accommodate the eight electrons originating from the ligands. One point to note in this bonding model is that symmetry allows some mixing of the metal p and d orbitals which share t_2 symmetry. The

Hybridization in a tetrahedral complex: in the valence bond model of coordination compounds, the formation of four bonds to a tetrahedral transition metal ion would involve four equivalent sp^3 hybrids.

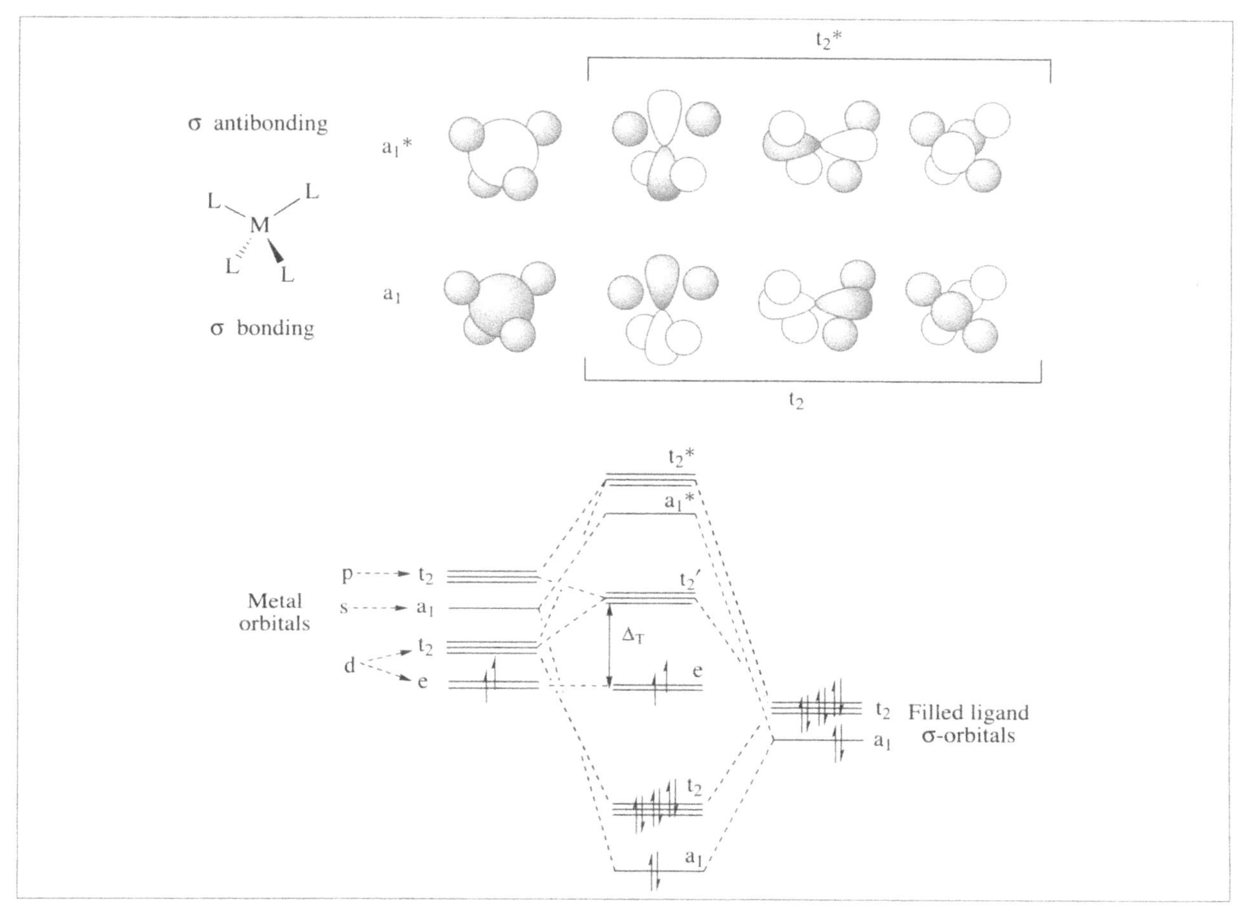

Figure 6.20 (a) σ bonding and σ* anti-bonding MOs in a tetrahedral complex, $[ML_4]$. (b) An MO energy level diagram for σ bonding in a tetrahedral complex

inclusion of π bonding in this model is more complicated than in the octahedral case. The eight π-type p orbitals of the ligands form eight LGOs of e, t_1 and t_2 symmetry. There are no metal valence shell orbitals of t_1 symmetry, so the t_1 π-LGOs give rise to a non-bonding MO. However, both the e and t_2 π-LGOs can interact with the e and t_2 symmetry orbitals of the σ bonding model to introduce new MOs (Figure 6.21). The effect of these interactions on the magnitude of Δ_T is not simple to predict, but it is important to note that the magnitude of Δ_T will reflect both π and σ bonding interactions.

The situation here is less straightforward than in the octahedral case, as both the e and the t_2' levels of the bonding MO scheme will be perturbed, and the relative effect on each cannot be predicted from a qualitative model. Interactions with π donors such as oxygen increase the energy of the e* and t_2'' MOs of the π bonding MO scheme and so favour higher oxidation states. Conversely, π acceptor ligands will reduce the energies of these orbitals, so favouring lower oxidation states.

The foregoing discussion shows how the crystal field model for bonding in transition metal complexes extracts, from the MO energy level diagram for σ bonding, those orbitals derived primarily from the metal valence shell d orbitals. To this extent the two models are consistent but,

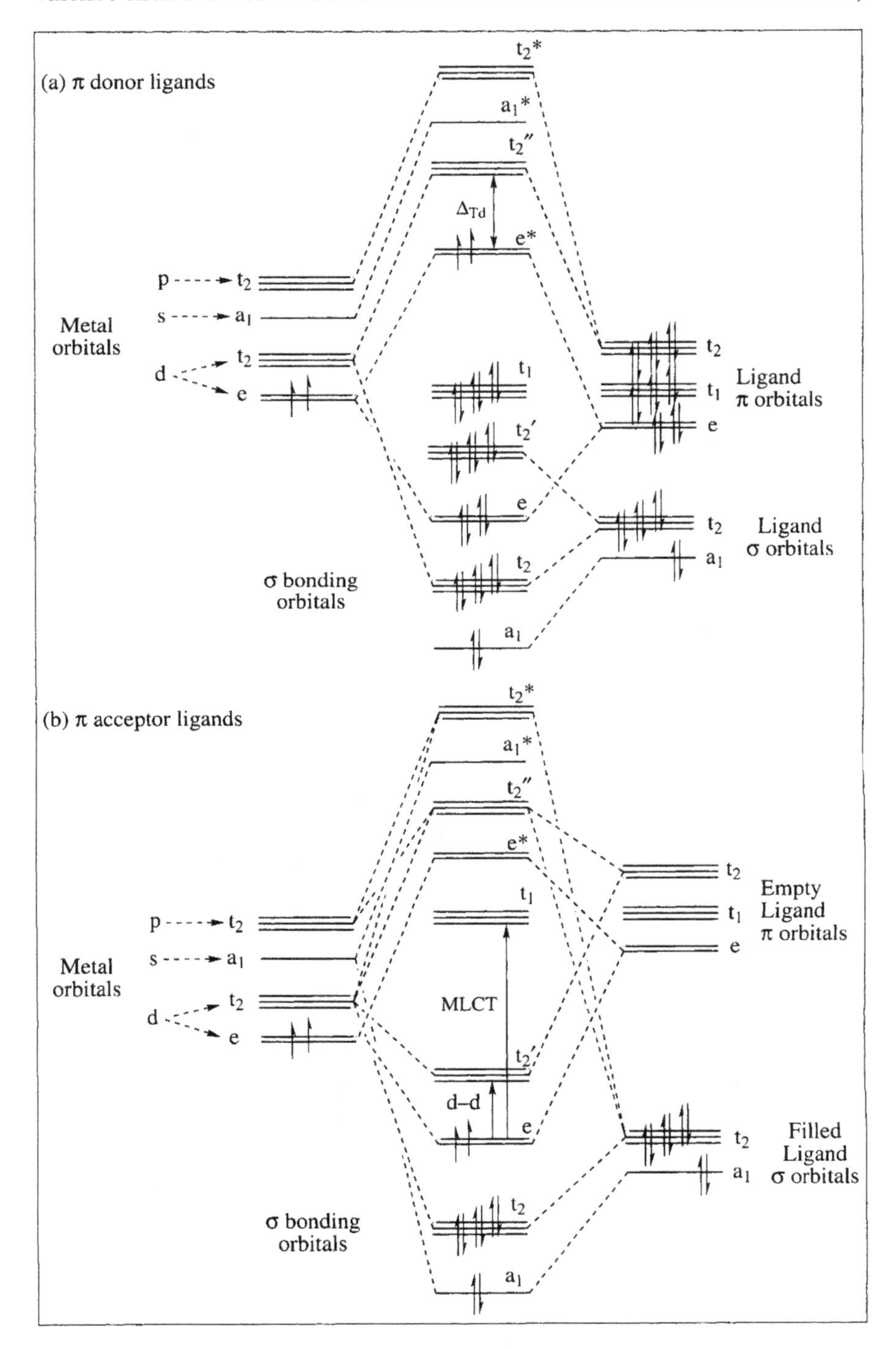

Figure 6.21 MO energy level diagrams including the effects of π bonding in a tetrahedral complex: (a) π donor ligands; (b) π acceptor ligands

although simpler, the crystal field theory does not naturally allow for π bonding, which can have a significant effect on the molecular orbital energy levels within a complex. Thus, although the crystal field theory is usually successful in predicting some of the major features of transition metal complexes, the MO model must be invoked to obtain a full understanding.

Summary of Key Points

1. *A simple ionic bonding model* accounts for many properties of transition metal complexes, including variations in the hydration and lattice enthalpies and the ionic radii of the metal ions. The observation of high- and low-spin states for complexes of some metal ions can also be explained.

2. *The bonding in tetrahedral and square planar complexes* can also be described using an ionic bonding model. The distortions found in the geometries of some metal ions can be explained by Jahn–Teller effects within the ionic model.

3. *A covalent bonding model is needed* to account for the effects of π-donor or π-acceptor ligands and synergic bonding. A molecular orbital description provides a more universal description of bonding within transtion metal complexes.

Problems

6.1. Develop a crystal field splitting diagram for the d orbitals of the metal in a trigonal bipyramidal complex $[ML_5]$ showing, qualitatively, how the energies of each of the d orbitals change.

6.2. (i) List five factors which may affect the magnitude of the the crystal field splitting in a transition metal complex.
(ii) Using the $10Dq$ values below, estimated from spectroscopic measurements, calculate the crystal field stabilization energies of the following complexes, in kJ mol^{-1} (assume a pairing energy of 19,000 cm^{-1} and that 1 kJ mol^{-1} = 83 cm^{-1}):

(a) $[Co(H_2O)_6]^{2+}$; $10Dq$ = 13,000 cm^{-1}.
(b) $[MnCl_6]^{4-}$; $10Dq$ = 15,000 cm^{-1}.
(c) $[CoCl_4]^{2-}$; assume $10Dq$ for octahedral $[CoCl_6]^{4-}$ is 21,000 cm^{-1}.

6.3. Using crystal field theory as a basis, explain why CN^- reacts with $[Fe(H_2O)_6]^{2+}$ to form $[Fe(CN)_6]^{4-}$, but with $[Ni(H_2O)_6]^{2+}$ to form $[Ni(CN)_4]^{2-}$.

6.4. Account for the pattern of Cu–E (E = O, N) bond distances found in the X-ray crystal structure of $[Cu\{OC(CF_3)CHC(CF_3)O\}_2(bpy)]$ (**6.1**), explaining fully the theoretical basis for your argument.

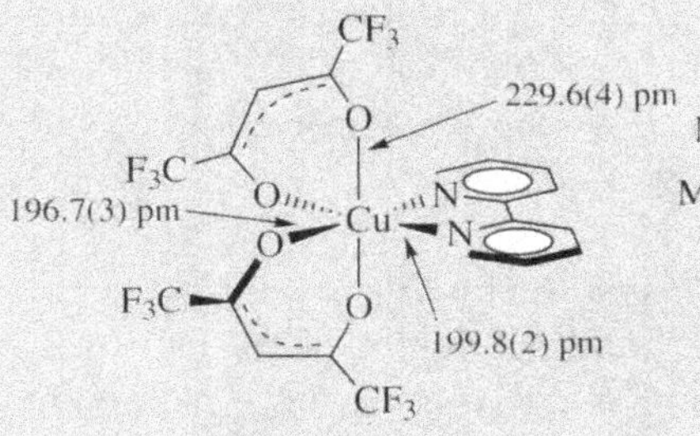

6.1

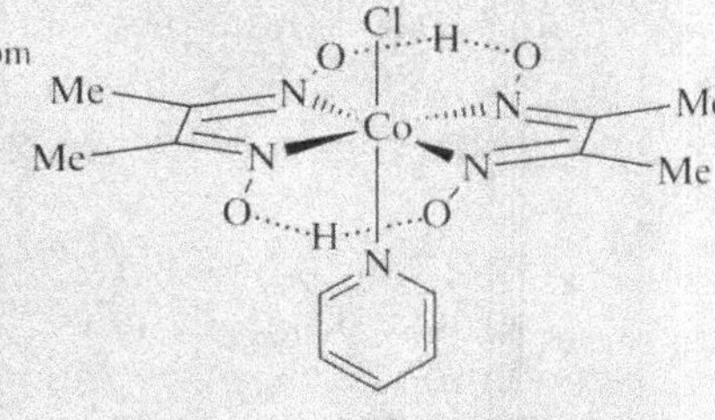

6.2

6.5. Use a crystal field theory model to predict the structural consequences of a two-electron reduction of the Co^{3+} complex **6.2**, which has no unpaired electrons, to give a product which also has no unpaired electrons. Explain fully the reasoning behind your answer, showing the relationship between structure and the energies of the individual d orbitals in the complexes.

6.6. Using an appropriate bonding description, explain the relative magnitudes of the crystal field splitting, $10Dq$ ($10Dq = \Delta_O$), in the following pairs of compounds:

(i)	$[CoF_6]^{3-}$	$10Dq = 13{,}100\ cm^{-1}$
	$[Co(NH_3)_6]^{3+}$	$10Dq = 22{,}900\ cm^{-1}$
(ii)	$[Fe(H_2O)_6]^{3+}$	$10Dq = 14{,}000\ cm^{-1}$
	$[Fe(CN)_6]^{3-}$	$10Dq = 32{,}000\ cm^{-1}$

6.7. (i) Explain what is meant by the term 'synergic bonding'.
(ii) Arrange the complexes **6.3**, **6.4** and **6.5** in order of decreasing C–O stretching frequency, explaining the reasons for your choice.

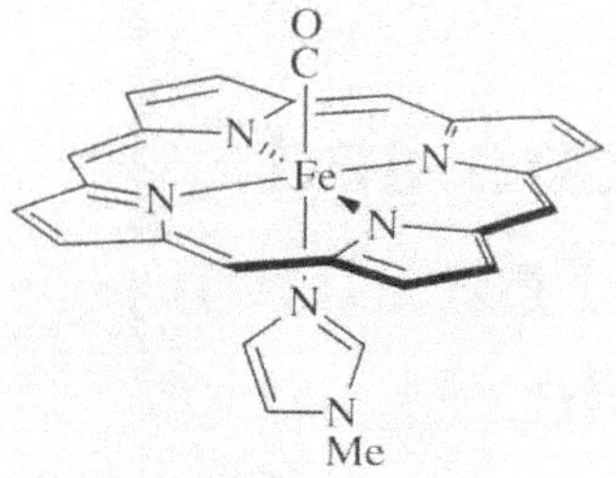

6.3

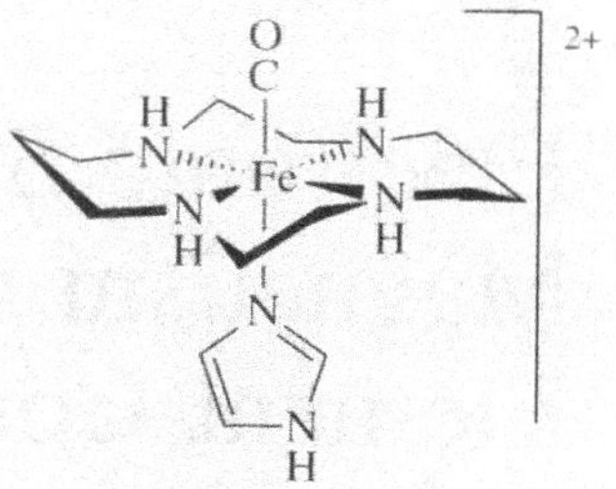

6.4

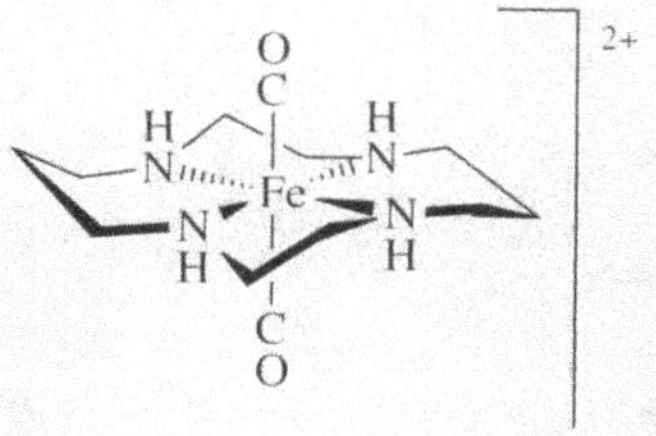

6.5

7

Electronic Spectra and Magnetism of Transition Element Complexes

Aims

After reading this chapter you should have gained an understanding of the origins of the electronic spectra and magnetism of transition element complexes and a knowledge of:

- The processes which lead to energy absorption in the electronic spectra of metal complexes
- The spectrochemical series of ligands
- The nephelauxetic effect
- The luminescence of lanthanide complexes
- Spin-only magnetic moments and their calculation

7.1 Introduction

One distinctive feature of transition element complexes is their varied colours. For example, aqueous solutions of $[Co(H_2O)_6]^{2+}$ ions are pale pink in colour, whereas solutions containing the $[CoCl_4]^{2-}$ ion are an intense blue colour. In order to explain such large differences in the colours of compounds of a particular metal ion, it is necessary to have some understanding of the origins of the light-absorbing processes. The colour which is observed in the visible spectrum complements the colour of the light which is absorbed by the complex. A blue colour corresponds to the absorption of light towards the red end of the visible spectrum and, conversely, a red colour corresponds with the absorption of light at the blue end of the spectrum. A plot of light absorbance against light energy for a complex produces its electronic spectrum and this will contain a series of absorption bands whose energies correspond with the energies of electronic transitions between particular MOs within the molecule. Thus the $[CoCl_4]^{2-}$ ion absorbs light of lower energy than that absorbed by the $[Co(H_2O)_6]^{2+}$ ion, reflecting differences in the electronic structures of the two complexes.

Another important feature of d- or f-block element compounds is their magnetism. A sample of green $[Ni(H_2O)_6]Cl_2$ is attracted into a magnetic field but, in contrast, a sample of the red complex $[Ni(dmgH)_2]$ ($dmgH_2$ = butane-2,3-dione dioxime, dimethylglyoxime) is weakly repelled by a magnetic field, yet both complexes contain Ni^{2+} ions. These magnetic effects arise from differences in the number of unpaired electrons in the complexes, so that simple magnetic observations can tell us something about the different electronic states of the Ni^{2+} ions in the two compounds. Thus the electronic spectra and magnetic properties of complexes provide an insight into their electronic structures.

7.2 The Electronic Spectra of Metal Complexes

The electronic spectrum of a metal complex provides a measure of the energy differences between the MOs which constitute the source and destination of the excited electron associated with each electronic transition. Since MOs have varying degrees of metal or ligand character, some electronic transitions may correspond with charge redistribution between the metal and the ligands whereas others may be primarily confined to ligand-based or to metal-based MOs. Four different types of electronic transition are possible within a metal complex.

7.2.1 Electronic Transitions

Intra-ligand Transitions

Transitions of this type are associated with polyatomic ligands which have electronic spectra in their own right. As an example, the tris(3,5-dimethylpyrazolyl)borate (tp^{Me_2-}) proligand (Figure 4.2) exhibits a $\pi \rightarrow \pi^*$ transition in the ultraviolet region at about 230 nm, which is also observed in its transition metal complexes. Usually, such absorptions do not reveal much about the metal centre and so they are of less interest to transition metal chemists than the remaining types of absorption which directly involve the metal centre.

Metal-to-ligand Charge Transfer

If light absorption causes an electron to be excited from an orbital based largely on the metal to an orbital based largely on the ligand, the absorption is described as a metal-to-ligand charge transfer (MLCT) band. It may be represented schematically by the process $M{-}L \rightarrow M^+{-}L^-$. In some complexes, transitions of this type can be exploited in the conversion of light energy into chemical energy through subsequent reactions of the reduced ligand.

Ligand-to-metal Charge Transfer

If light absorption causes an electron to be excited from an orbital based largely on the ligand to an orbital based largely on the metal, the absorption is described as a ligand-to-metal charge transfer (LMCT) band. It may be represented schematically by the process M–L → M^-–L^+. This type of electronic transition provides a basis for photography. The absorption of light by AgCl results in electron transfer from the Cl^- ion to the Ag^+ ion to give Ag metal, which creates the dark areas in a photographic negative.

Metal-based Electronic Transitions

In the case where light absorption leads to an electronic transition between orbitals which are largely metal in character, a d–d band will be observed for a d-block metal ion and an f–f transition for an f-block metal ion. In some cases, transitions between the *n*f and (*n* + 1)d orbitals are also observed in the spectra of f-block metal complexes. Not all of the possible electronic transitions in a metal complex are allowed to occur and selection rules govern which electronic transitions may be observed (Box 7.1).

Box 7.1 Selection Rules

Two important selection rules govern which electronic transitions of a metal complex may be observed. The first is the spin selection rule, which requires that there must be no change in the total spin quantum number *S* during the transition. As examples, singlet-to-singlet or triplet-to-triplet transitions are allowed, but triplet-to-singlet transitions are forbidden. In practice, spin-forbidden transitions may be observed, but will be very weak having typical molar absorptivities, ε, of <1 $dm^3\ mol^{-1}\ cm^{-1}$. A further selection rule, known as the Laporte rule, also applies to d–d or f–f transitions. This requires that, for ions in a centrosymmetric environment such as an octahedral field, the parity must change, g → u or u → g, as a result of the electronic transition. In an octahedral complex the d orbitals are all of g parity, so that d–d transitions are Laporte forbidden. In practice, vibrations occuring within the complex can reduce its symmetry on a timescale of *ca.* 10^{-13} s. Since the timescale for an electronic transition is much shorter, *ca.* 10^{-15} s, they are not totally prevented, and weak absorptions are seen with ε values of *ca.* 10 dm^3 mol^{-1} cm^{-1}. In tetrahedral complexes there is no centre of symmetry, so the Laporte rule is relaxed and ε values typically fall in the range 100 to 10^3 dm^3 mol^{-1} cm^{-1}. MLCT and LMCT processes are not confined to metal orbitals alone so are not Laporte forbidden. They are much more intense, with typical ε values in the range 10^3 to 10^5 dm^3 mol^{-1} cm^{-1}.

7.2.2 Electronic Spectra of Transition Metal Complexes

d–d Bands

In the electronic spectrum of a transition metal complex, d–d bands arise from electronic transitions which are largely localized on the metal ion. To a first approximation, therefore, d–d spectra can be analysed in terms of the states of a free metal ion subjected to the perturbation of the field produced by the ligand donor atom electrons. In a polyelectronic atom or ion the individual electrons each have spin and orbital angular momentum. However, when an observation is made of the atom, by electronic spectroscopy for example, it is the total resultant of all the electron spin and orbital angular momenta which is observed. Thus, in order to understand the spectroscopic and magnetic properties of metal ions, it is necessary to understand the ways in which the spin and orbital angular momenta of the individual electrons can combine to produce the properties of the ion as a whole, *i.e.* the properties which are actually observed. The spin and orbital angular momenta of an electron can be treated as independent quantities if the coupling between them is small. This generally applies to the first-row d-block metals, but not to the f-block elements where spin–orbit coupling becomes important. The starting point for analysing d–d spectra is the energy levels of the free metal ion, M^{z+}. This ion has closed shells of core electrons surrounded by a partly filled d subshell. When between two and eight d electrons are present, the different distributions of electrons among the d orbitals result in different combinations of spin and orbital angular momentum. These are specified by the total spin and orbital angular momentum quantum numbers, S and L, respectively. The values of S and L define the electronic state of the ion and, because of electron–electron interactions, not all arrangements of the electrons are of the same energy. The possible states for a d-block metal ion can be derived using the Russell–Saunders coupling scheme described in Box 7.2. This assumes that the coupling between electron spin and orbital angular momenta is small compared to coupling between electron spins, so that spin and orbital angular momenta may be treated separately. On this basis, term symbols may be assigned to the different distributions of electrons in the d subshell and the lowest energy or ground state predicted using Hund's rules (Box 7.3).

Box 7.2 The Russell–Saunders Coupling Scheme

In an atom or ion there are various different ways of arranging the electrons according to their spin and orbital angular momentum quantum numbers. This may be illustrated by considering a d^2 ion for which the orbital angular momentum quantum number l is 2, so each electron may adopt an m_l value of 2, 1, 0, –1 or –2 and an m_s value of +½ or –½. Thus the first electron may enter any of the five d orbitals with spin up or spin down, giving a total of 10 possibilities. As two electrons cannot have the same set of four quantum numbers, the second can be added in any of the nine remaining ways so there is a total of 90 possible ways to place two electrons in the five d orbitals. However, electrons are indistinguishable, so adding electron A then electron B is equivalent to adding B then A, so in the 90 ways each arrangement has been counted twice and there are only 45 distinct ways of placing two electrons in the five d orbitals. Each of these arrangements is known as a *microstate* (Figure 7.1). Because spin and orbital angular momenta are quantized vector quantities, a state with $S = 1$ has components with $M_S = 1$, 0 and –1, a state with $L = 1$ has components with $M_L = 1$, 0 and –1, and a state with $L = 2$ has components with $M_L = 2$, 1, 0, –1 and –2. This means that the 45 microstates must be collected into groups which reflect the quantization of S and L, as shown in Figure 7.1. Each of these groups constitutes a possible state of the d^2 system and can be given a symbol known as the *term symbol*. This is constructed from a super-prefix denoting the spin multiplicity ($2S + 1$) followed by a capital letter denoting the value of L as defined in Table 7.1:

$$^{(2S+1)}L$$

Hence the d^2 system can exist in one of the states ^1S, ^3P, ^1D, ^3F or ^1G. The states of other d electron configurations are summarized in Table 7.2.

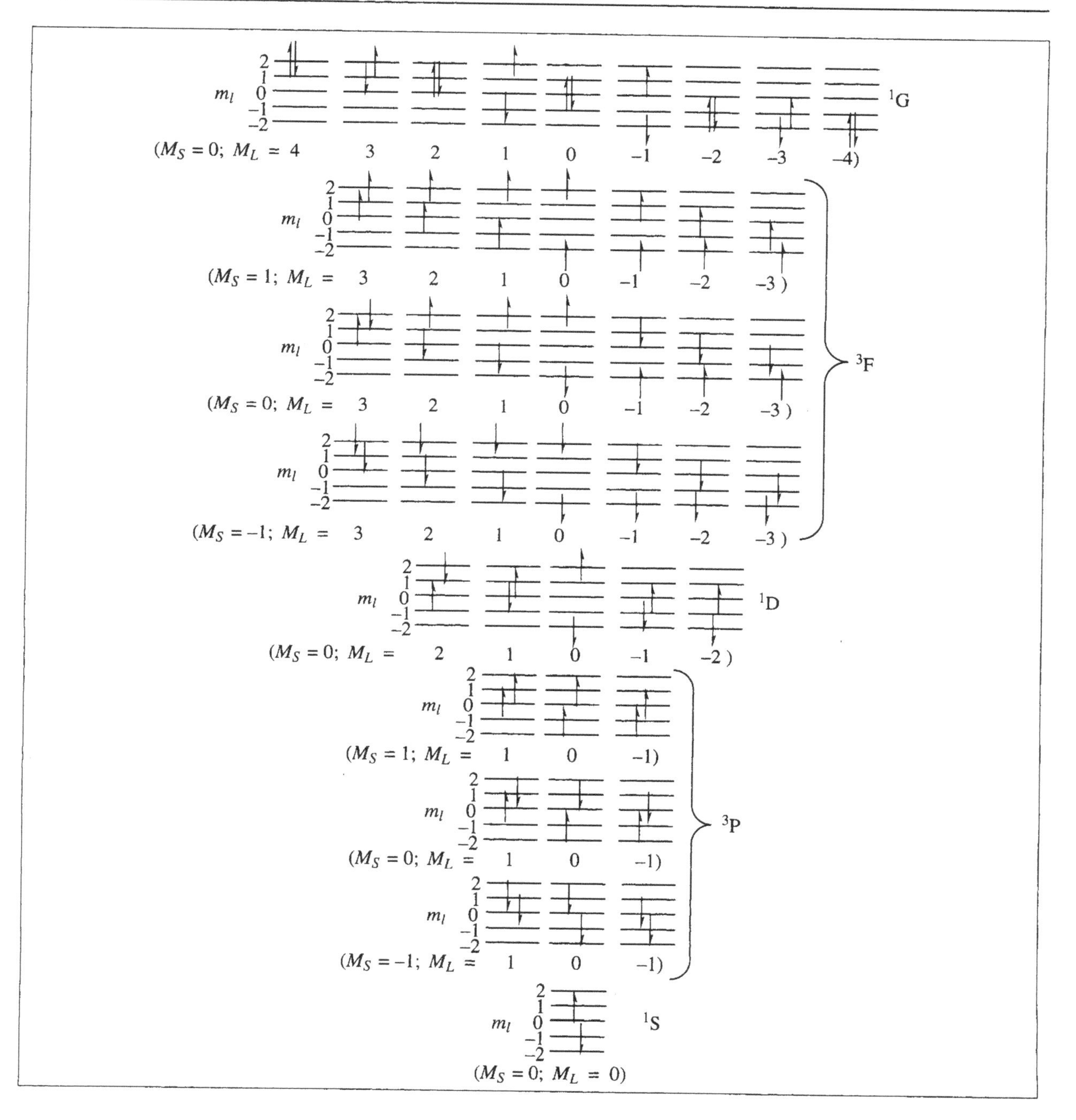

Figure 7.1 Microstates and Russell–Saunders terms for a d^2 transition metal ion

Table 7.1 Term symbols

L	0	1	2	3	4	5
Term symbol	S	P	D	F	G	H

Table 7.2 Russell–Saunders terms for d-block ions

Electron configuration[a]	*Russell–Saunders terms (ground term last)*
d^1, d^9	^2D
d^2, d^8	^1S, ^1D, ^1G, ^3P, ^3F
d^3, d^7	^2D, ^2P, ^2D, ^2F, ^2G, ^2H, ^4P, ^4F
d^4, d^6	^1S, ^1D, ^1G, ^1S, ^1D, ^1F, ^1G, ^1I, ^3P, ^3F, ^3P, ^3D, ^3F, ^3G, ^3H, ^5D
d^5	^2D, ^2P, ^2D, ^2F, ^2G, ^2H, ^2S, ^2D, ^2F, ^2G, ^2I, ^4P, ^4D, ^4F, ^4G, ^6S

[a]Because a single electron in an otherwise empty d shell behaves in the same way as an 'electron hole' in a filled d shell, the d^1 and d^9 configurations are equivalent with respect to their free ion terms. The other pairs of electron configurations shown are similarly related

Box 7.3 Hund's Rules

The energies of the various Russell–Saunders states of an atom or ion are not the same, and Hund's first and second rules can be used to decide which is the ground state:

(i) The lowest energy term (ground state) has the largest spin multiplicity ($2S + 1$).

(ii) If two terms have the same spin multiplicity, that with the larger value of L is of lowest energy.

These rules are usually accurate in defining the ground state but do not always predict the correct order for the higher energy terms, and for a d^2 ion the energy order of states is, starting from the ground state, $^3F < {}^1D < {}^3P < {}^1G < {}^1S$.

The simplest example for a d-block element is presented by a d^1 ion such as Ti^{3+}. Here the single electron has the quantum numbers $n = 3$, $l = 2$, $m_s = \pm\frac{1}{2}$ and $m_l = 2, 1, 0, -1$ or -2, so this electron configuration is assigned the term symbol ^2D. This is the ground term, in fact the only term, for a d^1 ion. In octahedral $[Ti(H_2O)_6]^{3+}$, the spectrum of which is shown in Figure 7.2, the degeneracy of the d orbitals is lifted by the ligand field which gives rise to the the t_{2g} and e_g* levels of the MO scheme

(Figure 6.15). The absorption of light by the complex at *ca.* 20,300 cm^{-1} (*ca.* 240 kJ mol^{-1}) corresponds with the promotion of the d electron from the t_{2g} to the e_g^* level, and so gives a direct measurement of Δ_O, or $10Dq$ (Figure 7.3). Molecular vibrations have the effect of broadening the absorption bands observed in electronic spectra, and this effect is apparent in d–d spectra. During a vibration of the metal–ligand bond the ligand–metal distances change, so the metal–ligand interaction changes. This gives rise to a range of energy differences between the orbitals associated with an electronic transition, for example between t_{2g} and e_g^* in $[Ti(H_2O)_6]^{3+}$ where Δ_O changes throughout the vibration. The spectroscopic measurement sums this effect over many molecules, leading to broad absorption bands.

In an octahedral field, the ^2D term of the free metal ion splits into two ligand field terms associated with the complex. When the single electron occupies the t_{2g} level the complexed ion is in its ground state $t_{2g}^1e_g^{*0}$, which is given the term symbol $^2T_{2g}$. When the single electron occupies the e_g^* level the complexed ion is in its first excited state $t_{2g}^0e_g^{*1}$, which is given the term symbol ^2E_g. Thus the electronic transition observed at 20,300 cm^{-1} for $[Ti(H_2O)_6]^{3+}$ corresponds with the excitation process $^2E_g \leftarrow {}^2T_{2g}$. The magnitude of Δ_O, or $10Dq$, depends on the nature of the

Representing electronic transitions: in accord with convention, the electronic transition from the $^2T_{2g}$ to the ^2E_g state is written with the ground state to the right. Although the example of a d^1 complex has been chosen for simplicity, it may not have escaped notice that both the $t_{2g}^1e_g^0$ and $t_{2g}^0e_g^1$ configurations should be subject to Jahn–Teller distortion (see Section 6.2.2). The lower energy shoulder in the absorption band of Figure 7.1 suggests that there are indeed more than just two energy levels involved here.

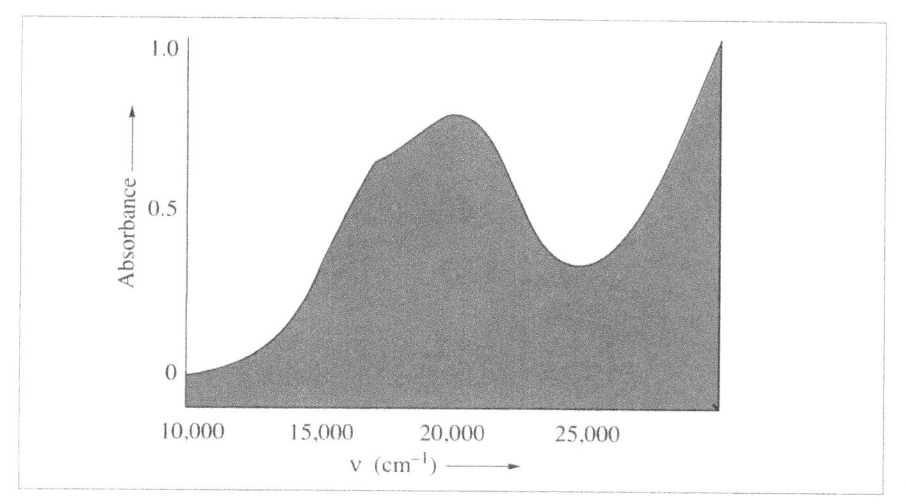

Figure 7.2 The electronic spectrum of $[Ti(H_2O)_6]^{3+}$

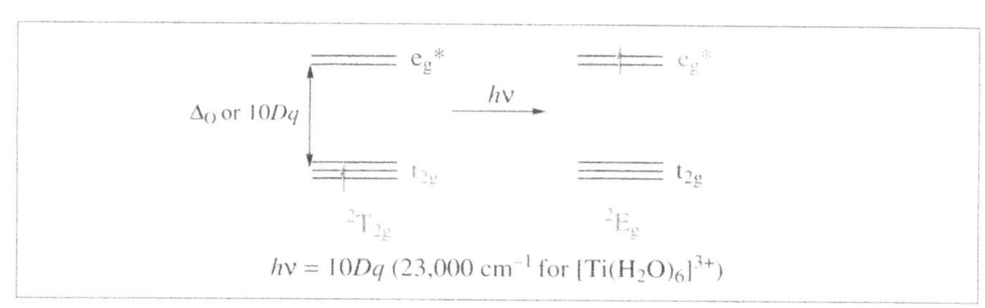

Figure 7.3 The electronic transition $t_{2g}^0e_g^1 \leftarrow t_{2g}^1e_g^0$

Two types of correlation diagrams have been constructed for octahedral and tetrahedral coordination geometries, and are named after the researchers who devised them. **Orgel diagrams** plot the energies of the ligand field terms of a d-block metal ion against ligand field parameters, and may show ligand field terms for tetrahedral and octahedral fields in the same diagram. A fixed zero level is used so that the energy of the ground term varies with the ligand field parameter *Dq* and only terms of the same spin multiplicity as the ground state are included. **Tanabe–Sugano diagrams** include all terms, and have the energy of the ground term set at zero so that this term constitutes the horizontal axis.

ligands bound to the metal ion, so it is possible to plot the energies of the $^2T_{2g}$ and ^2E_g terms as a function of ligand field strength in a correlation diagram as shown in Figure 7.4. Metal ions with more than one d electron give rise to more than one term (Table 7.2), each of which may be split by an octahedral ligand field as shown in Table 7.3. The situation is further complicated by electron–electron repulsion and whether this is large, small or of similar magnitude compared to the ligand field splitting parameter. Plots of the energies of the various ligand field terms of each metal electron configuration as a function of ligand field parameters have been made, and can be used to aid the interpretation of d–d spectra. In practice, although a number of free ion terms may be possible, the interpretation of an electronic spectrum is concerned with transitions from the ground state to higher energy states of the same spin multiplicity. Thus only a selected group of the possible terms will normally need to be considered. A simplified form of a diagram for the triplet terms of a d^2 ion, where the ligand field splitting is large compared to electron–electron repulsion terms, is shown in Figure 7.5. In this case the ground state is a triplet and, since transitions involving a change in *S* are not allowed, the ligand field terms derived from the free ion singlet terms have been omitted.

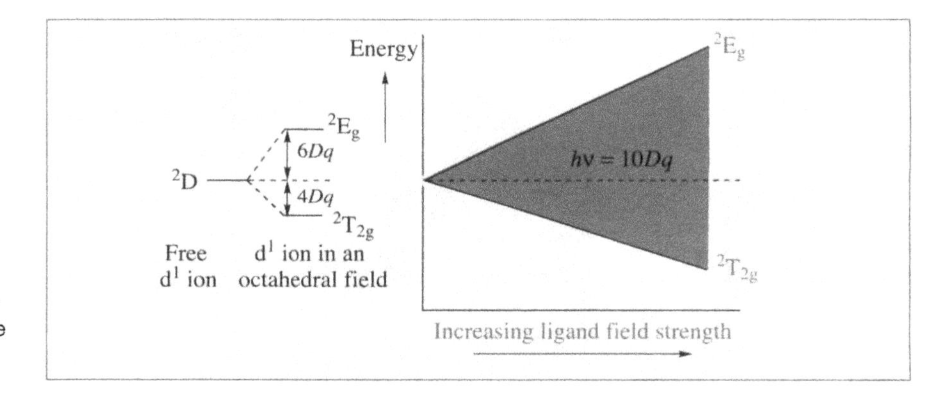

Figure 7.4 A correlation diagram showing the effect of octahedral ligand field strength on the energies of the $^2T_{2g}$ and ^2E_g terms of a d^1 metal ion

Table 7.3 Ligand field terms for an octahedral complex

Free ion term	*Number of states*	*Octahedral field terms*
S	1	A_{1g}
P	3	T_{1g}
D	5	T_{2g}, E_g
F	7	T_{1g}, T_{2g}, A_{2g}
G	9	A_{1g}, E_g, T_{1g}, T_{2g}

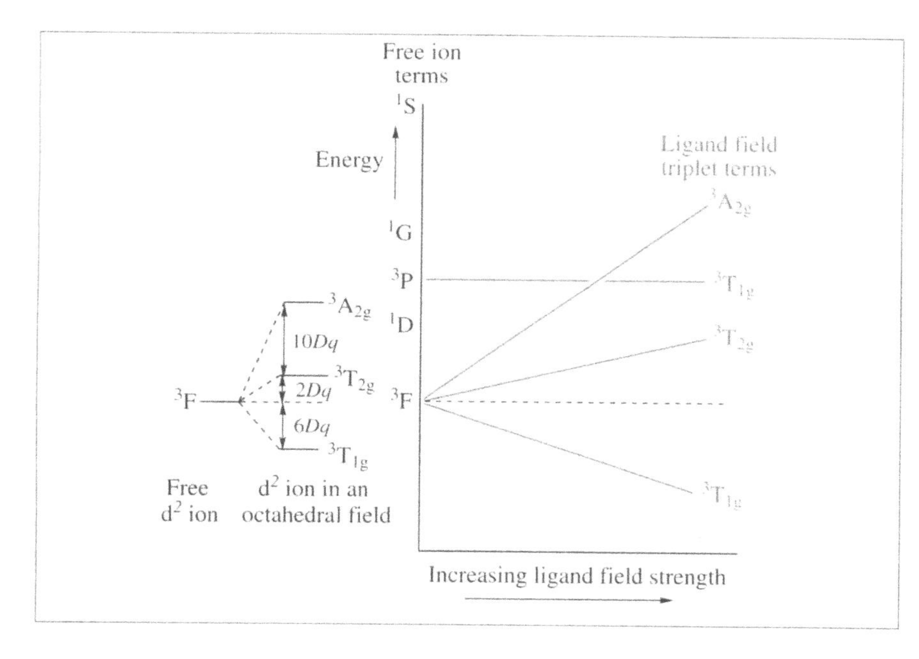

Figure 7.5 A correlation diagram showing the effect of octahedral ligand field strength on the energies of the triplet terms of a d^2 metal ion

Spectroscopic measurements of the ligand field splitting parameter, $10Dq$ or Δ_O, provide a means of ranking ligands in order according to the magnitude of splitting they induce in a given metal ion. This series is known as the spectrochemical series of ligands (see Section 6.2.6). The differences in the energies of the terms for a free metal ion are the result of electron–electron repulsions, and detailed spectroscopic measurements show that these repulsions are lower in metal complexes than in free ions. This reduced repulsion results from the delocalization of charge from the metal ion onto the ligands, thus expanding the region of space occupied by the electrons and reducing electron–electron repulsion. The term nephelauxetic (cloud expanding) effect has been used to describe this behaviour, and ligands can be placed in a nephelauxetic series according to their relative ability to delocalize charge as judged by a reduced electron–electron repulsion parameter. Larger, more polarizable, ligand donor atoms tend to be associated with increased electron delocalization, as do π acceptor ligands. The synergic bonding model of Section 6.3.2 provides a means of rationalizing these observations.

Worked Problem 7.1

Q Consult Figure 7.5 and assign the weak absorption bands in the electronic spectrum of $[V(H_2O)_6]^{3+}$ shown below, given that the $^3A_{2g} \leftarrow {}^3T_{1g}$ transition appears at 36,000 cm^{-1}. What is the value of $10Dq$ for $[V(H_2O)_6]^{3+}$?

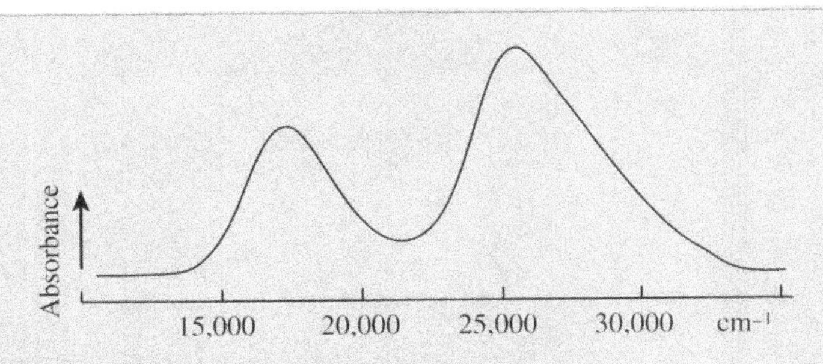

A The low ε values indicate that d–d transitions are involved. Based on Figure 7.5, three d–d transitions would be expected for an octahedral d^2 ion. These are $^3T_{2g} \leftarrow {}^3T_{1g}$, $^3T_{1g}(^3P) \leftarrow {}^3T_{1g}$ and $^3A_{2g} \leftarrow {}^3T_{1g}$. If $^3A_{2g} \leftarrow {}^3T_{1g}$ appears at high energy (36,000 cm^{-1}), the lowest energy band at *ca.* 17,500 cm^{-1} may be assigned to $^3T_{2g} \leftarrow {}^3T_{1g}$, and the intermediate energy band at *ca.* 26,000 cm^{-1} to $^3T_{1g}(^3P) \leftarrow {}^3T_{1g}$. Figure 7.5 indicates that, in this case, $10Dq$ is given by the difference in energy between the $^3T_{2g}$ and $^3A_{2g}$ terms, *i.e.* by the difference in energy between the two bands $^3T_{2g} \leftarrow {}^3T_{1g}$ and $^3A_{2g} \leftarrow {}^3T_{1g}$. This is *ca.* 36,000 – 17,500 = 18,500 cm^{-1}, a value similar in magnitude to that of 20,300 cm^{-1} found for $[Ti(H_2O)_6]^{3+}$.

Charge Transfer Bands

While d–d transitions arise from electronic transitions which are effectively localized on the metal ion, charge transfer processes involve the migration of charge between the metal ion and the ligands, and so need to be analysed with the aid of MO diagrams. LMCT processes involve the excitation of an electron from an orbital which is mainly ligand in character to one which is mainly metal in character. In an octahedral complex the electron must be excited into the t_{2g}* or e_g* MO (see Figure 6.17a), and so must originate from an occupied ground state orbital of u symmetry associated with the ligand to satisfy the Laporte selection rule. Since the ligand non-bonding σ orbitals are all of g symmetry, only the filled non-bonding ligand π orbitals of t_{1u} and t_{2u} symmetry might serve as the electron donor orbital. In LMCT spectra it might be expected that, as the metal ion becomes easier to reduce or the ligand easier to oxidize, the energy difference between the donor and acceptor orbitals will decrease and the LMCT absorption in the electronic spectrum will shift to lower frequency and longer wavelength. In $[IrBr_6]^{2-}$ (d^5) there are vacancies in both the t_{2g}* and e_g* levels, so that two charge transfer bands are observed at *ca.* 300 and 600 nm separated by aproximately Δ_O. In $[IrBr_6]^{3-}$ (low-spin d^6), t_{2g}* is filled so only the higher energy transition to e_g* is observed at *ca.* 300 nm.

In the example of a tetrahedral complex, LMCT involves excitation of an electron from the ligand non-bonding σ orbitals of a_1 and t_2 symmetry, or the ligand non-bonding π orbitals of e, t_1 or t_2 symmetry (see Figure 6.21a), to a predominantly metal-based e* or t_2'' orbital. In the case of $[MnO_4]^-$, the four lower-energy transitions observed in the UV/visible spectrum have been assigned to the transitions Mn(e*) $\leftarrow$ O(t_1), 17,700; Mn(t_2'') $\leftarrow$ O(t_1), 29,500; Mn(e*) $\leftarrow$ O(t_2), 30,300; and Mn(t_2'') $\leftarrow$ O(t_2), 44,400 cm^{-1}. The lowest energy transition at 17,700 cm^{-1} falls in the visible range and accounts for the purple colour of permanganate. In the series $[CuX_4]^{2-}$ (X = Cl, Br, I) the wavelength of the X to Cu LMCT band increases in the order Cl < Br < I. This corresponds with a decrease in the energy of the transition as the ligand field strength decreases and the energy of the Cu^{2+}-based t_2'' orbital falls relative to the halide-based donor orbital.

MLCT processes involve the excitation of an electron from an occupied orbital which is mainly metal in character to an unoccupied orbital which is mainly ligand in character. This requires the presence of π acceptor ligands such as CO or bipyridyl to provide empty ligand orbitals. In octahedral complexes the Laporte rule applies, and transitions from occupied t_{2g}, or possibly e_g*, orbitals to empty ligand t_{1u} or t_{2u} π* orbitals would give rise to MLCT bands (see Figure 6.17b). In a tetrahedral complex containing a π acceptor ligand, excitation of an electron from the predominantly metal-based e or t_2' orbitals to an unoccupied π* orbital of predominantly ligand character, such as t_1, would result in a migration of charge from the metal to the ligand (see Figure 6.21b).

7.2.3 Electronic Spectra of f-Block Ions

As for the d-block metal ions, f-block metal ion complexes may also show charge transfer bands in their electronic spectra, but, since the valence shell d orbitals are empty, no d–d transitions are observed. Instead, it is possible to observe f–f transitions and, in some cases, nf–(n + 1)d transitions which, for some actinides, are of low enough energy to appear in the ultraviolet region. The lack of significant crystal field effects in lanthanide ion complexes means that f–f spectra can usually be assigned simply on the basis of the spectroscopic terms of the free ion. However, the situation is complicated by the presence of significant spin–orbit coupling so that different J states have different energies (Box 7.4). The energy level diagram of an f^2 ion is shown in Figure 7.6 and may be applied to the Pr^{3+} ion, the spectrum of which is shown in Figure 7.7. The lack of significant crystal field effects mean that lanthanide f–f spectra show little vibrational broadening, so the absorption bands are narrow with typical bandwidths at half intensity of *ca.* 50 cm^{-1} (Figure 7.7). In the early actinides, spin–orbit coupling and crystal field effects are of similar magnitudes, making the interpretation of their electronic spectra more complicated. A study of bands due to 6d ← 5f transitions in uranium(+3) complexes has revealed a spectrochemical series:

	I^- <	Br^- <	Cl^- <	SO_4^{2-} <	H_2O ≈	F^-
ν(d←f)	13	17	19	22	25	25 × 10^3 cm^{-1}

Box 7.4 Lanthanide Ions

In the case of an f^2 configuration, the first electron can be added in any of 14 ways to the 4f orbitals and the second any of 13 ways, giving rise to (14 × 13)/2 = 91 microstates. Fortunately the term symbols for the possible states of these systems have been worked out already and can be taken as given, although it is important to appreciate the mechanism by which they were obtained. In energy order down to the ground state on the right these are:

$$^1S > {}^3P > {}^1I > {}^1D > {}^1G > {}^3F > {}^3H$$

Although the assumption that spin and orbital angular momenta do not couple is a good approximation in lighter elements, in heavier elements spin–orbit coupling becomes important. In these cases, S and L no longer adequately describe the state of the ion, and it is the resultant total spin and orbital angular momentum, j, for each electron which must be used. All the individual j values combine to give a resultant total angular momentum quantum number J through what is known as j–j coupling. The value of J can be related to the values of S and L obtained from the Russell–Saunders scheme by $J = S + L$. Since J is also quantized, resulting from the vectorial addition of S and L, J can take on values $|L + S|$, $|L + S| - 1$, *etc.*, to $|L - S|$. Some examples of Russell–Saunders states and the spin–orbit coupling J states which arise from them are given in Table 7.4. The third of Hund's rules defines which of the J states is lowest in energy:

(iii) If an electronic sub-shell is less than half filled, the state with the lowest value of J will be lowest in energy. If an electronic sub-shell is more than half filled, the state with the highest value of J will be lowest in energy.

The value of J is included in the term symbol as a post-subscript, so that the general expression for the term symbol is:

$$^{(2S+1)}L_J$$

where L represents the symbol for L defined in Table 7.1. The effects of the various couplings can be summarized in an energy level diagram for the f^2 case of the Pr^{3+} ion (Figure 7.6).

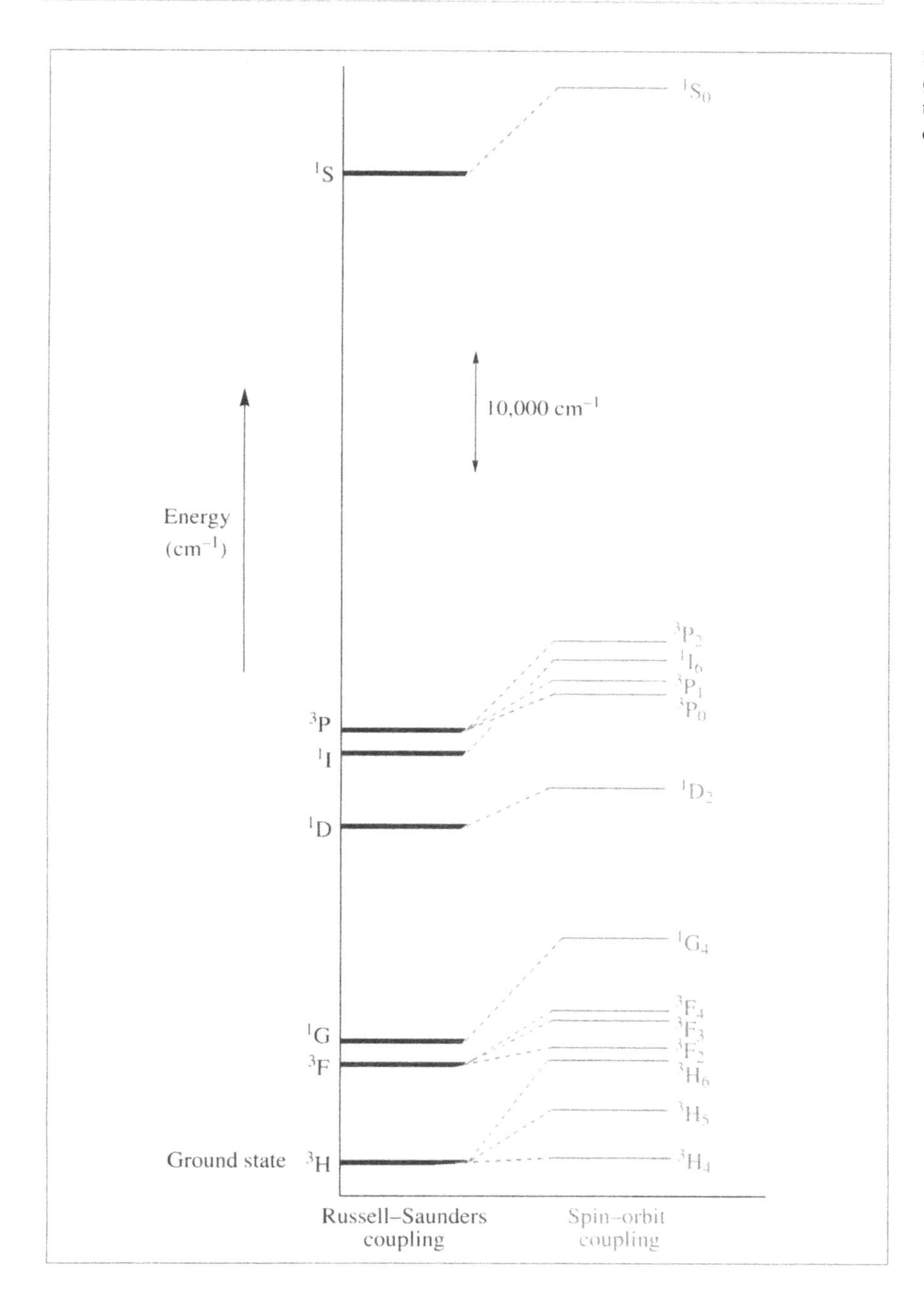

Figure 7.6 An energy level diagram for the f^2 ion Pr^{3+} showing the effect of spin–orbit coupling on the Russell–Saunders terms

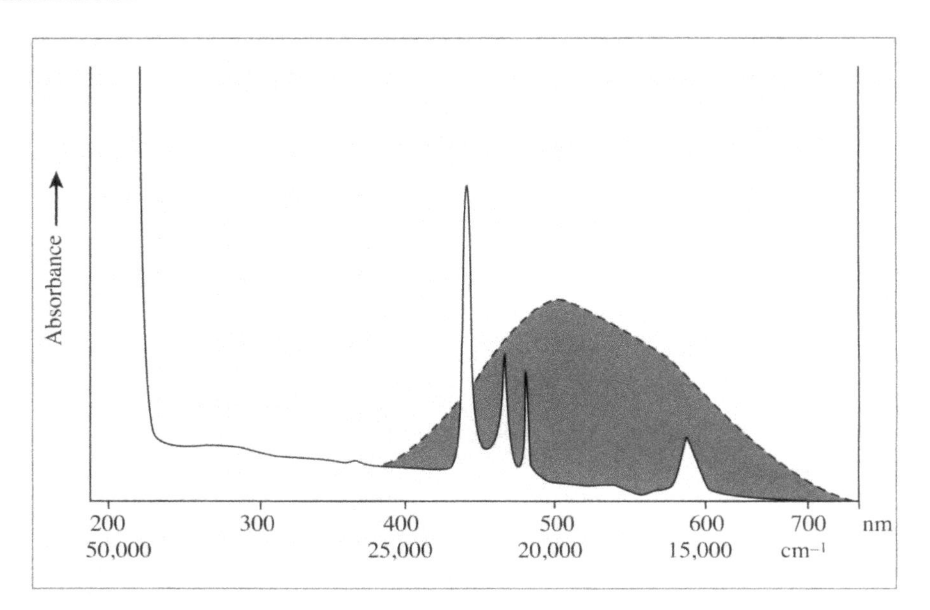

Figure 7.7 The electronic spectrum of Pr^{3+} (solid line) between 15,000 and 50,000 cm^{-1}. Several bands are observed in the low energy region 5,000 to 15,000 cm^{-1} which are not shown. The electronic spectrum of $[Ti(H_2O)_6]^{3+}$ is also shown (broken line) to illustrate the differing bandwidths of d-block and lanthanide ion spectra. Like d–d transitions, f–f transitions are Laporte forbidden and give weak absorptions

Table 7.4 Examples of terms resulting from spin–orbit coupling

Russell–Saunders term	*j–j coupling terms*
^1S	$^1S_0$
^3S	$^3S_1$
^2P	$^2P_{3/2}$, $^2P_{1/2}$
^3D	$^3D_3$, $^3D_2$, $^3D_1$
^7F	$^7F_6$, $^7F_5$, $^7F_4$, $^7F_3$, $^7F_2$, $^7F_1$, $^7F_0$
^2G	$^2G_{9/2}$, $^2G_{7/2}$

7.2.4 Lanthanide Luminescence

Ions which have energy gaps of suitable magnitude between excited and lower energy states, which are not bridged by non-radiative energy loss processes, may emit light as their excited states relax to a lower energy state. All Ln^{3+} ions except f^0 La^{3+} and f^{14} Lu^{3+} show some luminescent emission; Eu^{3+} and Tb^{3+} in particular show strong emissions. Direct excitation of Ln^{3+} ions is possible, but the low intensities and narrowness of the f–f absorption bands mean that intense radiation sources are required for effective excitation, *e.g.* lasers. A more widely available excitation mechanism is provided by the indirect excitation of Ln^{3+} through energy transfer from an excited state of a ligand which has a broad and intense intra-ligand absorption band (Box 7.5). Quenching of luminescence occurs when radiationless relaxation processes are fast enough to compete with the light emission process. This is a particular problem in

Perhaps the most commonly encountered example of **lanthanide luminescence** is found in colour cathode-ray TV tubes where red, blue and green phosphor dots excited by electron beams form the coloured picture elements. The respective red and green coloured phospors can contain Eu^{3+} in a Y_2O_2S or Y_2O_3 matrix and Tb^{3+} in a La_2O_2S matrix.

aqueous solutions, where energy transfer to overtones of O–H vibrations provides an efficient non-radiative relaxation pathway for the Ln^{3+} excited state. However, use of non-aqueous solvents and ligands which exclude water from the lanthanide coordination sphere provide conditions which allow luminescence to occur. The luminescent lifetimes of some Eu^{3+} and Tb^{3+} complexes can be as long as milliseconds, a feature which can be exploited in analytical applications where time-resolved measurements allow lanthanide luminescence to be observed after any shorter-lived fluorescence of a matrix containing the lanthanide ion has decayed.

Box 7.5 Lanthanide Ion Luminescence

The indirect excitation of a lanthanide ion initially involves excitation of a ligand electron to one of the vibrational levels of an excited state to form a ligand-centred excited singlet state, shown for the example of a Eu^{3+} complex in Figure 7.8. This rapidly relaxes to the lowest vibrational level of the excited state, and may undergo intersystem crossing to a lower energy triplet state from which energy transfer into one of the Ln^{3+} ion-based orbitals can occur. If relaxation of this excited state to a lower energy state or the ground state is accompanied by the emission of light, luminescence is observed.

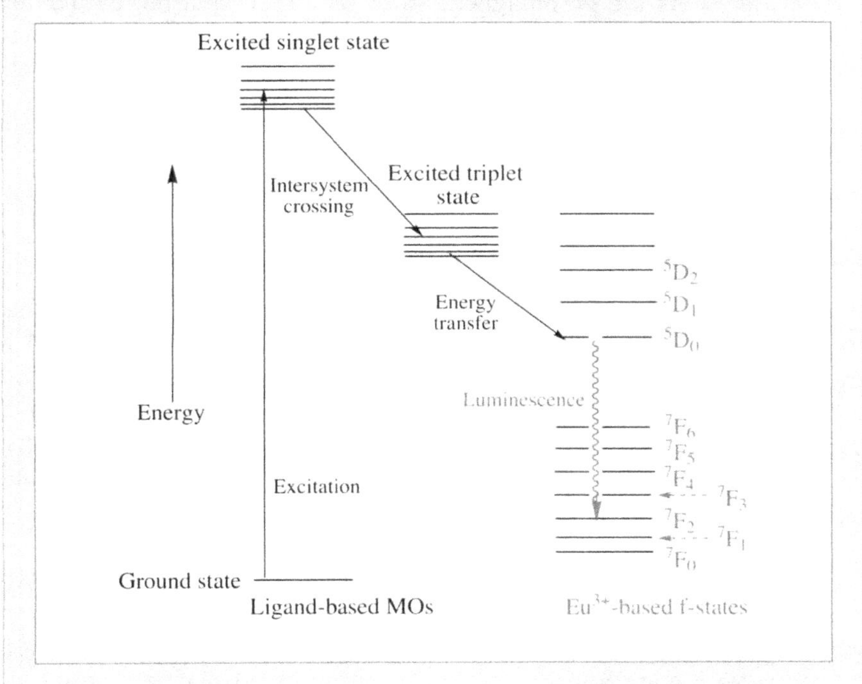

Figure 7.8 A schematic energy level diagram showing the origins of lanthanide luminescence through indirect excitation

7.3 The Magnetism of Transition Element Complexes

7.3.1 Magnetic Behaviour

There are two primary kinds of magnetic behaviour which may be observed when an atom, ion or molecule is placed in an applied magnetic field. The first is known as diamagnetism and arises from the circulation of electron pairs induced by the presence of the applied field. This charge circulation produces an induced magnetic moment which opposes the applied field, leading to the atom, ion or molecule being repelled by it. If unpaired electrons are present, the applied field also induces a second type of magnetic moment which is aligned with the magnetic field. This effect causes the sample to be attracted into the field, and is known as paramagnetism. Paramagnetic effects are typically some 10 to 10^4 times larger than diamagnetic effects so that, when present, paramagnetism usually dominates the observed magnetic behaviour of the sample. Paramagnetism results from the interaction of the spin and orbital angular momentum of electrons with the applied magnetic field. Assuming that there is no coupling between the spin and orbital angular momenta of a metal ion, the magnetic moment, μ_{SL}, of an atom or ion is related to the total spin quantum number S and the total orbital angular momentum quantum number L by equation 7.1. In this equation, g represents the gyromagnetic ratio or Landé splitting parameter which, for a free electron, is 2.00023. The units of μ_{SL} are Bohr magnetons (BM).

$$\mu_{SL} = \sqrt{g^2S(S+1) + L(L+1)} \tag{7.1}$$

7.3.2 Magnetism of Transition Metal Ions

In practice, spin–orbit coupling is small for first-row d-block metal ions, so that equation 7.1 may be applied. Furthermore, the d orbitals of a transition metal ion usually show significant interactions with the ligand orbitals. In such cases, the orbital angular momentum contribution is usually quenched so that $gS(S + 1) >> L(L + 1)$. Where this happens, equation 7.1 can be simplified to the spin-only formula shown in equation 7.2:

$$\mu_S = g\sqrt{S(S+1)} \tag{7.2}$$

where μ_S is known as the spin-only magnetic moment. Often, equation 7.2 provides a good approximation to the observed magnetic moment,

μ_{obs}, of a first-row transition metal complex, and generally it is found that $\mu_S \leq \mu_{obs} \leq \mu_{SL}$, as shown by Table 7.5. In the case of high-spin d^5 ions such as Mn^{2+} or Fe^{3+}, $L = 0$, so that in these cases $\mu_{obs} = \mu_S$. The crystal field model, and a knowledge of whether a complex is high or low spin, allows the number of unpaired electrons in a complex to be calculated, and so its magnetic moment may be predicted. The magnetic behaviour of paramagnetic second- and third-row d-block metal ions is less simple, as spin–orbit coupling and the effects of covalency in bonding are more pronounced.

Table 7.5 Examples of the magnetic moments of some transition metal complexes (μ in BM)

Metal ion	*S*	*L*	μ_{SL}[a]	μ_{obs}	μ_S
V^{4+}	1/2	2	3.00	1.7–1.8	1.73
V^{3+}	2/2	3	4.47	2.6–2.8	2.83
Cr^{3+}	3/2	3	5.20	~3.8	3.87
Co^{3+}	4/2	2	5.48	~5.4	4.90
Fe^{3+}[b]	5/2	0	5.92	~5.9	5.92

[a]See Worked Problem 7.2 for examples of calculating μ_{SL}
[b]Note that, for the half-filled shell, $L = 0$, so $\mu_{obs} = \mu_S$

Magnetic measurements can be used to obtain information about the structures adopted by metal complexes. One example is provided by the reversible binding of dioxygen to the protein haemoglobin, which contains Fe^{2+} and transports oxygen around the body. On binding O_2, the iron centre switches from a high-spin state with four unpaired electrons to a low-spin diamagnetic state. This information has been used as a basis for distinguishing between possible bonding models for the Fe^{2+}–O_2 interaction, which is chemically unusual because of its reversible nature. Normally, Fe^{2+} is readily and irreversibly oxidized to Fe^{3+} by dioxygen.

Worked Problem 7.2

Q (i) Calculate the magnetic moments, μ_{SL}, of Cr^{3+} and Co^{3+} ions in the absence of a crystal field.

(ii) Calculate the spin-only magnetic moments of the Ni^{2+} ion in $[Ni(H_2O)_6]^{2+}$ and the Fe^{2+} ion in high-spin $[Fe(H_2O)_6]^{2+}$.

A (i) The magnetic moment of a metal ion may be calculated from its S and L values according to equation 7.1 as shown in Figure 7.9.

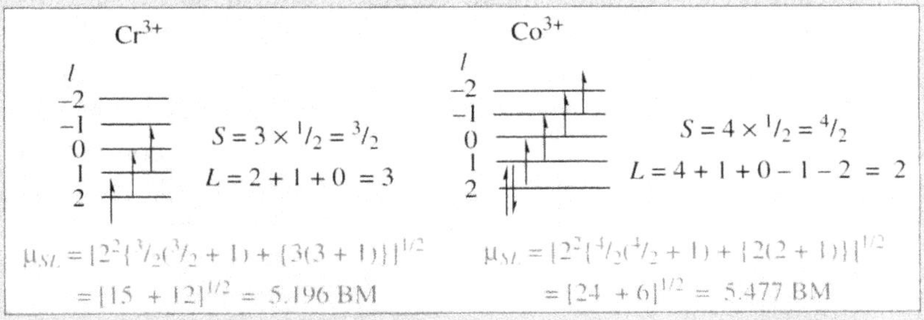

Cr^{3+}

$S = 3 \times {}^1/_2 = {}^3/_2$

$L = 2 + 1 + 0 = 3$

$\mu_{SL} = [2^2\{{}^3/_2({}^3/_2 + 1) + \{3(3 + 1)\}]^{1/2}$
$= [15 + 12]^{1/2} = 5.196$ BM

Co^{3+}

$S = 4 \times {}^1/_2 = {}^4/_2$

$L = 4 + 1 + 0 - 1 - 2 = 2$

$\mu_{SL} = [2^2\{{}^4/_2({}^4/_2 + 1) + \{2(2 + 1)\}]^{1/2}$
$= [24 + 6]^{1/2} = 5.477$ BM

Figure 7.9 Examples of calculating the magnetic moments of metal ions for d^3 Cr^{3+} and high-spin d^6 Co^{3+}.

(ii) The spin-only formula may be applied to these cases as follows:

$[Ni(H_2O)_6]^{2+}$ contains Ni^{2+}, a d^8 ion. This gives an octahedral structure, $t_{2g}^6e_g^2$ with two unpaired electrons, so:

$$\mu_S = 2\sqrt{{}^2\!/_2({}^2\!/_2 + 1)} \quad = 2.83 \text{ BM}$$

$[Fe(H_2O)_6]^{2+}$ contains octahedral high-spin Fe^{2+} which is d^6 with the electron configuration $t_{2g}^4e_g^2$, giving four unpaired electrons, so:

$$\mu_S = 2\sqrt{{}^4\!/_2({}^4\!/_2 + 1)} \quad = 4.90 \text{ BM}$$

7.3.3 The Magnetism of Lanthanide Ions

The magnetism of f-block metal ions is complicated by the fact that spin–orbit coupling can be of comparable magnitude to, or larger than, the crystal field splitting. Its effects cannot be ignored, and equation 7.1 does not apply. In such cases it is the total angular momentum quantum number, J, which must be used, where J represents the sum of the individual combined spin and orbital angular momenta, j, of the individual electrons. In such cases the magnetic moment μ_J is given by equation 7.3, in which the value of g_J is given by equation 7.4:

$$\mu_J = g_J\sqrt{J(J+1)} \tag{7.3}$$

$$g_J = {}^3\!/_2 + \frac{S(S+1) - L(L+1)}{2J(J+1)} \tag{7.4}$$

The absence of significant crystal field effects in the lanthanide ions means that the number of unpaired electrons can be simply predicted from the electron configuration of the ion. This value varies from one to seven for f^1 to f^7 configurations, then from six to one for f^8 to f^{13} configurations, with Lu^{3+} being a closed-shell diamagnetic ion. In practice, equation 7.3 predicts the observed magnetic moments of tripositive lanthanide ion complexes quite well, as shown by Figure 7.10. The two discrepancies at Sm^{3+} and Eu^{3+} arise because the first excited state of these ions is sufficiently close to the ground state that, at room temperature, thermal energy is sufficient to partly populate the excited state. Since the J value of the first excited state differs from that in the ground state, the observed magnetic moment is no longer purely that of the ground state and the excited state magnetic moment becomes mixed into the value of μ_{obs}. If allowance is made for this effect, the magnetic moments of the Eu^{3+} and Sm^{3+} ions can also be successfully predicted.

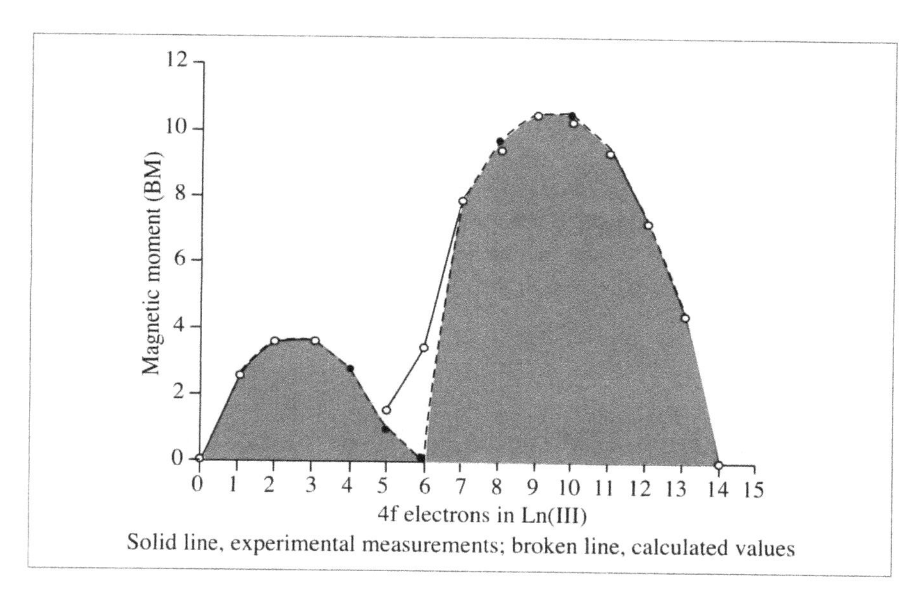

Figure 7.10 A plot of the magnetic moments of Ln^{3+} ions against the number of f electrons

Worked Problem 7.3

Q Calculate the magnetic moment of the Nd^{3+} ion, with the ground state $^4I_{9/2}$.

A The Nd^{3+} ion has a $4f^3$ electron configuration, and the ground state term symbol $^4I_{9/2}$ shows that $S = 3/2$, $L = 6$ and $J = 9/2$. The value of g_J for Nd^{3+} can be obtained using equation 7.4 and the value of μ_J can then be obtained using equation 7.3:

$$g_J = \frac{3}{2} + \frac{\frac{3}{2}(\frac{3}{2}+1) - 6(6+1)}{2\{\frac{9}{2}(\frac{9}{2}+1)\}} = \frac{3}{2} - \frac{38.25}{49.5} = 0.7273$$

$$\mu_J = g_J\sqrt{\tfrac{9}{2}(\tfrac{9}{2}+1)} = 3.62 \text{ BM to 3 significant figures}$$

Summary of Key Points

1. *Any of four electronic excitation processes can lead to light absorption* in the visible or ultraviolet region of the spectrum of metal complexes.

2. *Electronic spectral absorptions* associated with electronic transitions between d orbitals provide information about the magnitude of crystal field splitting. Anomalies in the results obtained indicate the need for a more covalent bonding model to describe some complexes.

3. *Transitions between higher and lower energy f orbitals* in excited state lanthanide ions lead to luminescence properties.

4. *The magnetic properties of d- or f-block metal ion complexes* can usually be predicted from ionic bonding models. Orbital contributions to magnetism can often be neglected for first row d-block metal ions but must be included when considering f-block metal ions.

Problems

(Assume 1 kJ mol^{-1} corresponds to 83 cm^{-1})

7.1. Describe the four processes which can lead to the absorption of light by a transition metal complex.

7.2. The electronic spectrum of a complex $[TiL_6]^{3+}$ (L is a neutral monodentate ligand) shows a weak ($\varepsilon = 7$ dm^3 mol^{-1} cm^{-1}) absorption maximum at an energy of 19,200 cm^{-1}. Explain the origin of this absorption. Given that the pairing energy for Fe^{3+} is 260 kJ mol^{-1}, calculate the magnetic moment for a complex $[FeL_6]^{3+}$, assuming that the magnitude of the crystal field splitting of Fe^{3+} with L is similar to that with Ti^{3+}.

7.3. The electonic spectrum of a Ti^{3+} complex $[TiX_6]^{3-}$ (X is a π donor ligand) contains a weak ($\varepsilon = 10$ dm^3 mol^{-1} cm^{-1}) absorption band at 19,000 cm^{-1} with a shoulder and four intense ($\varepsilon = 10^4$ dm^3 mol^{-1} cm^{-1}) bands at 8000, 12,000, 27,000 and 31,000 cm^{-1}. Explain the origins of these spectroscopic features.

7.4. Determine the geometries and spin-only magnetic moments of the following complexes, where X represents a neutral monodentate weak-field ligand and Z a neutral monodentate strong-field ligand; explain your reasoning:
(i) $[FeX_4]^{3+}$; (ii) $[FeZ_6]^{2+}$; (iii) $[NiZ_4]^{2+}$; (iv) $[NiX_6]^{2+}$; (v) $[CuX_6]^{2+}$.

7.5. Assuming a pairing energy of 16,000 cm^{-1}, calculate the crystal field stabilization energies (in kJ mol^{-1}) and spin-only magnetic moments of the following complexes:

(i) $[Fe(H_2O)_6]^{2+}$; assume $10Dq$ = 14,000 cm^{-1}.
(ii) $[Fe(CN)_6]^{3-}$; assume $10Dq$ = 24,000 cm^{-1}.
(iii) $[MnCl_4]^{2-}$; assume $10Dq$ = 19,000 cm^{-1} for the corresponding octahedral complex.
(iv) $[CoCl_4]^{2-}$; assume $10Dq$ = 17,000 cm^{-1} for the corresponding octahedral complex.

7.6. Calculate the magnetic moment of the Dy^{3+} ion with ground state $^6H_{15/2}$.

Further Reading

A short introductory text such as this cannot hope properly to convey the breadth and depth of chemical knowledge about the transition elements. A variety of texts provide further information about the topics introduced here, and should be consulted during your studies to expand your chemical knowledge and your understanding of the topic. Some provide a body of descriptive chemistry,[1–9] while others approach the subject more through the underlying principles.[10–20] Chemistry primers providing an introduction to d-block and f-block chemistry are available,[21–24] while more specialized accounts of some of the topics discussed here are given in other texts.[4,8,17–20] In particular, texts on group theory will be useful for those wishing to pursue further the topics in Chapters 6 and 7.[25–28] Several websites now provide information on the transition elements and their chemistry.[29–31]

References

1. N. N. Greenwood and A. Earnshaw, *Chemistry of the Elements*, 2nd edn., Butterworth-Heinemann, Oxford, 1997.
2. F. A. Cotton and G. Wilkinson, *Advanced Inorganic Chemistry*, 5th edn., Wiley, New York, 1988.
3. C. E. Housecroft and A. G. Sharpe, *Inorganic Chemistry*, Prentice Hall, Harlow, 2001.
4. S. A. Cotton, *Lanthanides and Actinides*, Macmillan, London, 1991.
5. M. N. Hughes, *The Inorganic Chemistry of Biological Processes*, 2nd edn., Wiley, Chichester, 1981.
6. *Comprehensive Coordination Chemistry*, ed. R. D. Gillard, J. A. McCleverty and G. Wilkinson, Pergamon Press, Oxford, 1987.
7. *Kirk-Othmer Encyclopaedia of Chemical Technology*, ed. J. I. Kroschwitz and M. Howe-Grant, Wiley, New York, 1996.

8. S. J. Lippard and J. M. Berg, *Principles of Bioinorganic Chemistry*, University Science Books, Mill Valley, CA, 1994.
9. J. D. Lee, *Concise Inorganic Chemistry*, 5th edn., Blackwell, Oxford, 1996.
10. W. W. Porterfield, *Inorganic Chemistry, A Unified Approach*, 2nd edn., Academic Press, San Diego, 1993.
11. B. E. Douglas, D. H. McDaniel and J. J. Alexander, *Concepts and Models of Inorganic Chemistry*, Wiley, New York, 1994.
12. D. F. Shriver and P. W. Atkins, *Inorganic Chemistry*, 3rd edn., Oxford University Press, Oxford, 1999.
13. J. E. Huheey, E. A. Keiter and R. L. Keiter, *Inorganic Chemistry*, 4th edn., Harper & Row, Cambridge, 1993.
14. U. Müller, *Inorganic Chemistry: A Co-ordination Chemistry Approach*, Wiley, New York, 1992.
15. G. L. Meissler and D. A. Tarr, *Inorganic Chemistry*, 2nd edn., Prentice Hall, Englewood Cliffs, NJ, 1999.
16. D. M. P. Mingos, *Essential Trends in Inorganic Chemistry*, Oxford University Press, Oxford, 1998.
17. M. Gerlock and E. C. Constable, *Transition Metal Chemistry – the Valence Shell in d-Block Chemistry*, VCH, Weinheim, 1994.
18. J. Barrett, *Understanding Inorganic Chemistry, The Underlying Physical Principles*, Ellis Horwood, Chichester, 1991.
19. S. F. A. Kettle, *Physical Inorganic Chemistry: A Co-ordination Chemistry Approach*, Spektrum, Oxford, 1996.
20. E. C. Constable, *Metals and Ligand Reactivity*, VCH, Weinheim, 1996.
21. M. J. Winter, *d-Block Chemistry*, Oxford Science Publications, Oxford, 1994.
22. J. A. McCleverty, *Chemistry of the First Row Transition Metals*, Oxford Science Publications, Oxford, 1999.
23. D. E. Fenton, *Bioco-ordination Chemistry*, Oxford Science Publications, Oxford, 1995.
24. N. Kaltsoyannis and P. Scott, *The f Elements*, Oxford Science Publications, Oxford, 1999.
25. M. Ladd, *Symmetry and Group Theory in Chemistry*, Ellis Horwood, Chichester, 1999.
26. A. Vincent, *Molecular Symmetry and Group Theory*, Wiley, Chichester, 1977
27. G. Davidson, *Group Theory for Chemists*, Macmillan, Basingstoke, 1991.
28. F. A. Cotton, *Chemical Applications of Group Theory*, Wiley, New York, 1990.

29. Information relating to the Periodic Table and the properties of elements, including the d- and f-block, may be found at the website of Dr M. J. Winter: http://www.webelements.com/

30. Information relating to the chemistry of the f-block elements may be found at the website of Dr S. J. Heyes: http://www.chem.ox.ac.uk/icl/heyes/LanthAct/lanthact.html

Answers to Problems

Chapter 1

1.1. (i) Reduction of the Au^{3+} in $[AuCl_4]^-$ to gold metal.
(ii) Removal of chloride as insoluble $PbCl_2$ to allow silver to dissolve as $AgNO_3$.
(iii) The acid sulfate salt is able to selectively dissolve the rhodium as Rh^{3+} sulfate, which is soluble, while leaving the remaining metals in elemental form.
(iv) The peroxide is able to oxidize the remaining metals to produce oxide derivatives, from which IrO_2 can be separated after dissolution of the ruthenium and osmium compounds in water.
(v) The hydroxide is used to trap the volatile neutral osmium tetroxide in a soluble anionic compound of low volatility.

1.2. (i) Examples are provided by:
(a) Dioxygen transport by haemoglobin or haemerythrin and dioxygen storage in myoglobin.
(b) Oxidation of substrates catalysed by peroxidase enzymes, *e.g.* equation A.1:

$$SubH_2 + H_2O_2 \rightarrow Sub + 2H_2O \quad \text{catalysed by Fe(peroxidase)} \quad (A.1)$$
$$(SubH_2 = \text{substrate})$$

(c) Disproportionation of peroxide by catalase enzymes (equation A.2):

$$2H_2O_2 \rightarrow 2H_2O + O_2 \quad \text{catalysed by Fe(catalase)} \quad (A.2)$$

(d) Electron transport by cytochromes based on the Fe^{3+}/Fe^{2+} couple.
(ii) Examples are provided by:

(a) The hydration of carbon dioxide by carbonic anhydrase (equation A.3):

$$CO_2 + 2H_2O \rightleftharpoons HCO_3^- + H_3O^+ \quad (A.3)$$

(b) The hydrolysis of proteins by carboxypeptidase (equation A.4):

$$\text{Protein-CONHCH(R)}CO_2^- + H_2O \rightarrow \text{Protein-}CO_2^- + {}^+NH_3CH(R)CO_2^- \quad (A.4)$$

(R = an amino acid side chain)

(c) The disproportionation of superoxide to O_2 and O_2^{2-} by superoxide dismutase (SOD) containing copper and zinc (equations A.5 and A.6):

$$O_2^- + Cu^{2+}/Zn^{2+}(SOD) \rightarrow O_2 + Cu^+/Zn^{2+}(SOD) \quad (A.5)$$

$$O_2^- + Cu^+/Zn^{2+}(SOD) \rightarrow O_2^{2-} + Cu^{2+}/Zn^{2+}(SOD) \quad (A.6)$$

(iii) Examples are provided by:
(a) Electron transport by copper blue proteins based on the Cu^{2+}/Cu^+ couple.
(b) The transport of dioxygen by haemocyanin.
(c) The reduction of dioxygen by cytochrome *c* oxidase [Fe^{2+}/Cu^+(CytOx)] containing copper and iron (equation A.7):

$$O_2 + Fe^{2+}/Cu^+(CytOx) \rightarrow O_2^{2-} + Fe^{3+}/Cu^{2+}(CytOx) \quad (A.7)$$

1.3. Zinc cannot undergo electron transfer reactions, so would be unsuitable in any process involving a change in metal valency. Also, Zn^{2+} is a filled d subshell ion and is chemically different from Fe^{2+} or Cu^{2+}, which have partly filled d subshells.

1.4. (i) Primary +2, secondary 4.
(ii) Primary +2, secondary 6.
(iii) Primary +4, secondary 6.
(iv) Primary +3, secondary 6.
(v) Primary +2, secondary 4.

Chapter 2

2.1. The first two IEs involve removal of the 4s electrons, which are shielded from the increasing nuclear charge by the 3d electrons and so are relatively independent of Z. The second IE of copper is anomalously high because Cu^+ has the electron configuration $3d^{10}4s^0$ with an associated loss of exchange energy in forming Cu^{2+}. The third IE involves the removal of a 3d electron from the M^{2+} ion, in which 4s is empty. Because 3d electrons do not shield one another from the nuclear charge efficiently, the IE values for this ionization are more sensitive to increasing Z. This gives rise to an underlying trend of IE increasing with Z. Added to this, there is an increasing exchange energy (Box 2.1) loss in ionizing 3d electrons. This reaches a maximum at $3d^5$ Mn^{2+}, after which there is a discontinuity at $3d^6$ Fe^{2+}, which has one electron pair. Ionizing this does not incur any loss in exchange energy, so the IE plot is displaced to lower values. Following this displacement, the trend seen in the first half of the row is repeated to $3d^{10}$ Zn^{2+}, which has five electron pairs, and suffers the maximum loss of exchange energy on the third ionization.

2.2. Valence shell electron configurations:

Cu^+ (+1) $3d^{10}$, TaO_4^{3-} (+5) $5d^0$, Pr^{3+} (+3) $4f^2$, OsO_4 (+8) $5d^0$, Rh^+ (+1) $4d^8$, Gd^{3+} (+3) $4f^7$, Yb^{2+} (+2) $4f^{14}$, $PaO(OH)^{2+}$ (+5) $5f^0$, UO_2^{2+} (+6) $5f^0$, Pu^{4+} (+4) $5f^4$, No^{2+} (+2) $5f^{14}$.

2.3. Two trends are apparent in Table 2.8. Firstly, in going from Sc^{3+} to Y^{3+} to La^{3+} there is a stepwise increase in radius as principal shells of electrons are added. However, in all of the other groups the increase between the first and second row is not repeated in the second to third row. This is a result of the lanthanide contraction and the filling of the 4f subshell between La and Hf. This reflects the poor screening of 4f electrons one by another, leading to an increase in Z_{eff} and a decrease in radius. Secondly, for metals in the same oxidation state, there is a d-block contraction across the rows as a result of the increase in Z_{eff}.

2.4. The relatively poor shielding of one 3d electron by another leads to an increasing Z_{eff} experienced by a further electron added to a first-row d-block metal atom. There is also an increasing exchange energy benefit until the half-filled $3d^5$ subshell is reached at manganese. Here, adding an electron involves creating an electron pair and a consequence of this is a positive E^a value for manganese. Following this discontinuity the trend of the first part of the period repeats with the magnitude of the electron gain energy, increasing to copper where $3d^{10}4s^2$ Cu^- is being formed. The value of E^a then becomes positive at zinc, where the formation of Zn^- involves adding an electron to the 4p subshell beyond the filled $3d^{10}4s^2$ configuration.

2.5. (i) Higher oxidation states are favoured towards the left of the transition metal series where IEs are lower. Since Mn^{2+} is d^5, there will be a maximum loss of exchange energy on breaking into the half-filled subshell to form d^4 Mn^{3+}. Thus the third IE of manganese is higher than that of chromium, making Mn^{3+} thermodynamically less accessible. In the case of Fe^{2+}, oxidation to Fe^{3+} involves going from d^6 to d^5 with no loss of exchange energy. This leads to a lower third IE, facilitating oxidation to Fe^{3+}.
(ii) The one-electron reduction of dioxygen to superoxide is thermodynamically difficult and cannot be effected by Fe^{2+}. It is easier to perform the two-electron reduction to peroxide and, since Fe^{2+} is a one-electron reductant, the interaction of two Fe^{2+} ions with one dioxygen molecule is necessary to form peroxide. In an aqueous solution of Fe^{2+} there is nothing to prevent this and the reaction proceeds. However, in haemoglobin the Fe^{2+} site is protected by the encapsulating protein, so that only a 1:1 Fe^{2+}:O_2 complex can form. Under these circumstances, Fe^{2+} is unable to reduce dioxygen and the binding becomes reversible.

2.6. (i) In the lanthanides the f orbitals are core-like and there is insufficient covalency in the bonding to support the formation of higher oxidation states beyond +3. In the early actinides the the 5f orbitals are less core-like. In addition, the 5f and 6d orbitals are more similar in energy, so that d orbital participation can increase the covalent contribution to bonding, allowing oxidation states above +4 to form. Hence UO_2^{2+} is stable in aqueous solution. However, as the atomic number increases, the 5f orbitals become more core-like, the IEs increase, bonding becomes more ionic, and oxidation state +3 becomes more dominant.
(ii) This is probably a result of the interplay between ionization, vaporization and bond formation enthalpies. The ionization enthalpies of the later actinides are not all known with certainty, but redox potentials for the $An^{4+/3+}$ and $An^{3+/2+}$ couples show a steady increase from Cm to No, showing that oxidation state +3 becomes harder to achieve. The same is true for Yb in that the sum

of the first three IEs is increasing at Yb, although this is compensated by the decreased vaporization energy of Yb. The third IE of No appears to be larger than that of Yb, and it is possible that the smaller size of Yb^{3+} results in a larger bond formation energy, making Yb^{3+} more accessible than its counterpart No^{3+}.

Chapter 3

3.1. The general form of the plot of atomization enthalpies is the inverse of the plot of the third IEs. The more difficult the third ionization, the lower the atomization enthalpy. A consideration of IEs and physical properties among the heavier elements shows that high IEs are associated with lower boiling or melting points, *e.g.* the halogens, the Group 18 gases, Hg and Cd. A consideration of the solution to Worked Problem 2.4 indicates that, among the lanthanides, metal–metal distances will be at a maximum for Eu and Yb. Also the high third IEs of Eu and Yb mean that Ln^{2+} ions are present in a conduction band containing only two electrons per atom. This reduces the energy needed to remove a metal atom from the bulk, and gives a reduced enthalpy of atomization.

3.2. The calculations for MnO are as follows:

$$\Delta H_U^o = -\frac{6.022\times10^{23}\times1.748\times2\times2\times(1.602\times10^{-19})^2}{4\times3.142\times8.854\times10^{-12}\times(83+140)\times10^{-12}}\times(1-\tfrac{1}{8})$$

$$= -\frac{108.1\times10^{-15}}{24815\times10^{-24}}\times0.875 = -3812 \text{ kJ mol}^{-1}$$

The thermochemical cycle value of the lattice energy for MnO is quoted as 3745 kJ mol^{-1} so that our calculated value is less than 2% different from this.

For MnO_2:

$$\Delta H_U^o = -\frac{6.022\times10^{23}\times2.408\times4\times2\times(1.602\times10^{-19})^2}{4\times3.142\times8.854\times10^{-12}\times(53+136)\times10^{-12}}\times(1-\tfrac{1}{8})$$

$$= -\frac{297.7\times10^{-15}}{21031\times10^{-24}}\times0.875 = -12,386 \text{ kJ mol}^{-1}$$

3.3. The calculations are shown in Scheme A.1.

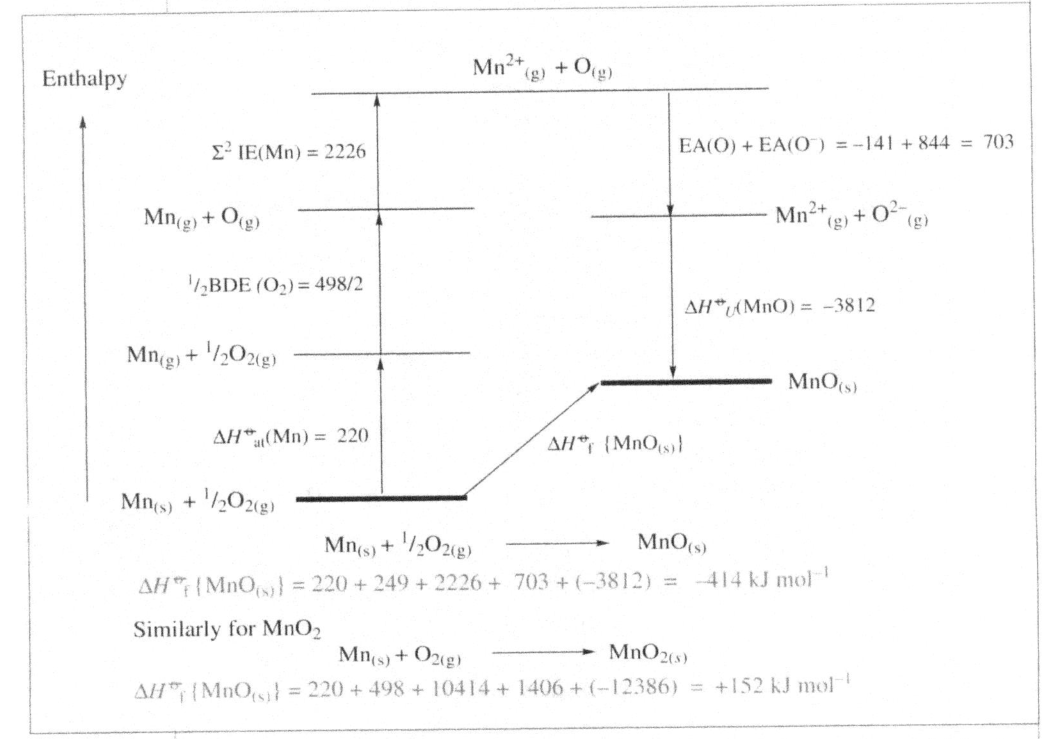

Scheme A.1

The experimental values for the enthalpies of formation of MnO and MnO_2 are respectively −385 and −529 kJ mol^{-1}. The value for MnO is in good agreement with that of −414 kJ mol^{-1} obtained using the calculated lattice energy. However, the experimental value for MnO_2 is significantly more exothermic than the calculated value of +152 kJ mol^{-1}. This may reflect the presence of an additional contribution to bonding from covalency, which is not included in the ionic Born–Landé equation model.

3.4. The temperature T^d at which the Gibbs free energy of the disproportionation reaction of VCl_3, shown in equation 3.10, becomes zero is given by $\Delta H^\ominus/\Delta S^\ominus$:

$$\Delta H^\ominus = -452 + (-569) - 2(-581) = +141 \text{ kJ mol}^{-1}$$
$$\Delta S^\ominus = 97 + 255 - 2(131) = +90 \text{ J K}^{-1} \text{ mol}^{-1}$$
$$\Delta H^\ominus/\Delta S^\ominus = T^d = 141000/90 = 1567 \text{ K or } 1294\ ^\circ\text{C}.$$

According to this calculation, above *ca.* 1294 °C the free energy for reaction 3.10 becomes negative and the reaction becomes spontaneous.

3.5. The solution to this problem involves applying the principles illustrated in the previous question to assess the temperatures at which the various chlorides decompose or disproportionate according to equations A.8–A.12:

$$2UCl_{6(s)} \rightarrow U_2Cl_{10(s)} + Cl_{2(g)} \quad (A.8)$$
$$\Delta H^\ominus = -2192 + 0 - 2(-1138) = +84 \text{ kJ mol}^{-1}$$
$$\Delta S^\ominus = 470 + 222 - 2(285) = +122 \text{ J K}^{-1} \text{ mol}^{-1}$$
$$\Delta H^\ominus/\Delta S^\ominus = T^d = 84000/122 = 689 \text{ K or } 416\ °\text{C}.$$

$$U_2Cl_{10(s)} \rightarrow UCl_{4(s)} + UCl_{6(s)} \quad (A.9)$$
$$\Delta H^\ominus = -1050 + (-1138) - (-2192) = +4 \text{ kJ mol}^{-1}$$
$$\Delta S^\ominus = 197 + 285 - 470 = +12 \text{ J K}^{-1} \text{ mol}^{-1}$$
$$\Delta H^\ominus/\Delta S^\ominus = T^d = 4000/12 = 333 \text{ K or } 60\ °\text{C}.$$

$$U_2Cl_{10(s)} \rightarrow 2UCl_{4(s)} + Cl_{2(g)} \quad (A.10)$$
$$\Delta H^\ominus = -2(1050) + 0 - (-2192) = +92 \text{ kJ mol}^{-1}$$
$$\Delta S^\ominus = 2(197) + 222 - 470 = +146 \text{ J K}^{-1} \text{ mol}^{-1}$$
$$\Delta H^\ominus/\Delta S^\ominus = T^d = 92000/146 = 630 \text{ K or } 357\ °\text{C}.$$

$$4UCl_{4(s)} \rightarrow U_2Cl_{10(s)} + 2UCl_{3(s)} \quad (A.11)$$
$$\Delta H^\ominus = -2192 + 2(-891) - 4(-1050) = +226 \text{ kJ mol}^{-1}$$
$$\Delta S^\ominus = 470 + 2(165) - 4(197) = +12 \text{ J K}^{-1} \text{ mol}^{-1}$$
$$\Delta H^\ominus/\Delta S^\ominus = T^d = 226000/12 = 18{,}833 \text{ K or } 18{,}560\ °\text{C}.$$

$$UCl_{4(s)} \rightarrow UCl_{3(s)} + \tfrac{1}{2}Cl_{2(g)} \quad (A.12)$$
$$\Delta H^\ominus = -891 + 0 - (-1050) = +159 \text{ kJ mol}^{-1}$$
$$\Delta S^\ominus = 165 + 222/2 - (197) = +79 \text{ J K}^{-1} \text{ mol}^{-1}$$
$$\Delta H^\ominus/\Delta S^\ominus = T^d = 159000/79 = 2013 \text{ K or } 1740\ °\text{C}.$$

These calculations show that, on heating, UCl_6 can lose chlorine to form U_2Cl_{10} (T^d = 416 °C), which readily disproportionates to UCl_4 and UCl_6 ($\Delta G^\ominus = 0$ at *ca.* 60 °C). In addition, U_2Cl_{10} can lose chlorine at the decomposition temperature of UCl_6 to form UCl_4, which is thermally stable with a T^d of 1740 °C. The negative enthalpy of formation, and positive entropy, of UCl_3 indicate that the compound is thermally stable. However, it is very readily oxidized.

3.6. To answer this question it is necessary to find whether or not $\Delta G^\ominus$ for the reaction in equation A.13 is negative at 1073 K:

$$MnO_{2(s)} + V_2O_{4(s)} \rightarrow MnO_{(s)} + V_2O_{5(s)} \qquad (A.13)$$
$$\Delta H^{\ominus} = -385 + (-1560) - (-521) - (-1439) = +15 \text{ kJ mol}^{-1}$$
$$\Delta S^{\ominus} = 60 + 131 - 53 - 103 = +35 \text{ J K}^{-1} \text{ mol}^{-1}$$
$$\Delta G^{\ominus} = \Delta H^{\ominus} - T\Delta S^{\ominus} = 15 - 1073 \times 35/1000 = -22.6 \text{ kJ mol}^{-1}$$

Although the reaction is endothermic, above 156 °C (T^{d}) the entropy term $-T\Delta S^{\ominus}$ becomes sufficiently large to overcome $\Delta H^{\ominus}$ and provide a driving force for the reaction. At 800 °C the reaction is quite exergonic with $\Delta G^{\ominus} = -22.6$ kJ mol^{-1}.

Chapter 4

4.1. Some possible answers are shown in Figure A.1 opposite.

4.2. Seven types of isomerism with examples are:

(i) Geometric, *e.g.* square planar or tetrahedral; *cis* or *trans*; *mer* or *fac*:
cis-$[Pt(NH_3)_2Cl_2]$ or *trans*-$[Pt(NH_3)_2Cl_2]$, *mer*-$[Co(NH_3)_3Cl_3]$ or *fac*-$[Co(NH_3)_3Cl_3]$.

(ii) Linkage, *e.g.* M–NCS or M–SCN in $[Pd(SCN)_2(bpy)]$ or $[Pd(NCS)_2(bpy)]$; M–NO2 or M–ONO in red $[Co(NH_3)_5(NO_2)]Cl_2$ or yellow $[Co(NH_3)_5(ONO)]Cl_2$.

(iii) Ionization, *e.g.* $[Co(NH_3)_5Br]SO_4$ or $[Co(NH_3)_5(SO_4)]Br$.

(iv) Coordination, *e.g.* $[Co(NH_3)_6][Cr(CN)_6]$ or $[Cr(NH_3)_6][Co(CN)_6]$.

(v) Hydrate (or solvate), *e.g.* $[Cr(H_2O)_4Cl_2]Cl.2H_2O$ or $[Cr(H_2O)_6]Cl_3$.

(vi) Polymerization, *e.g.* $[Pt(NH_3)_2Cl_2]$ or $[Pt(NH_3)_4][PtCl_4]$.

(vii) Optical, *e.g.* enantiomers of $[MoCl(tp^{Me}{}_2)(O)(OPh)]$ (**A.1**) and $[Co(en)_3]^{3+}$ (**A.2**).

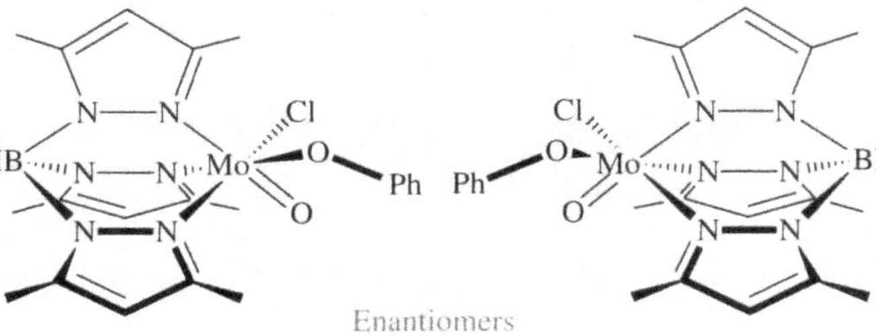

Enantiomers

A.1

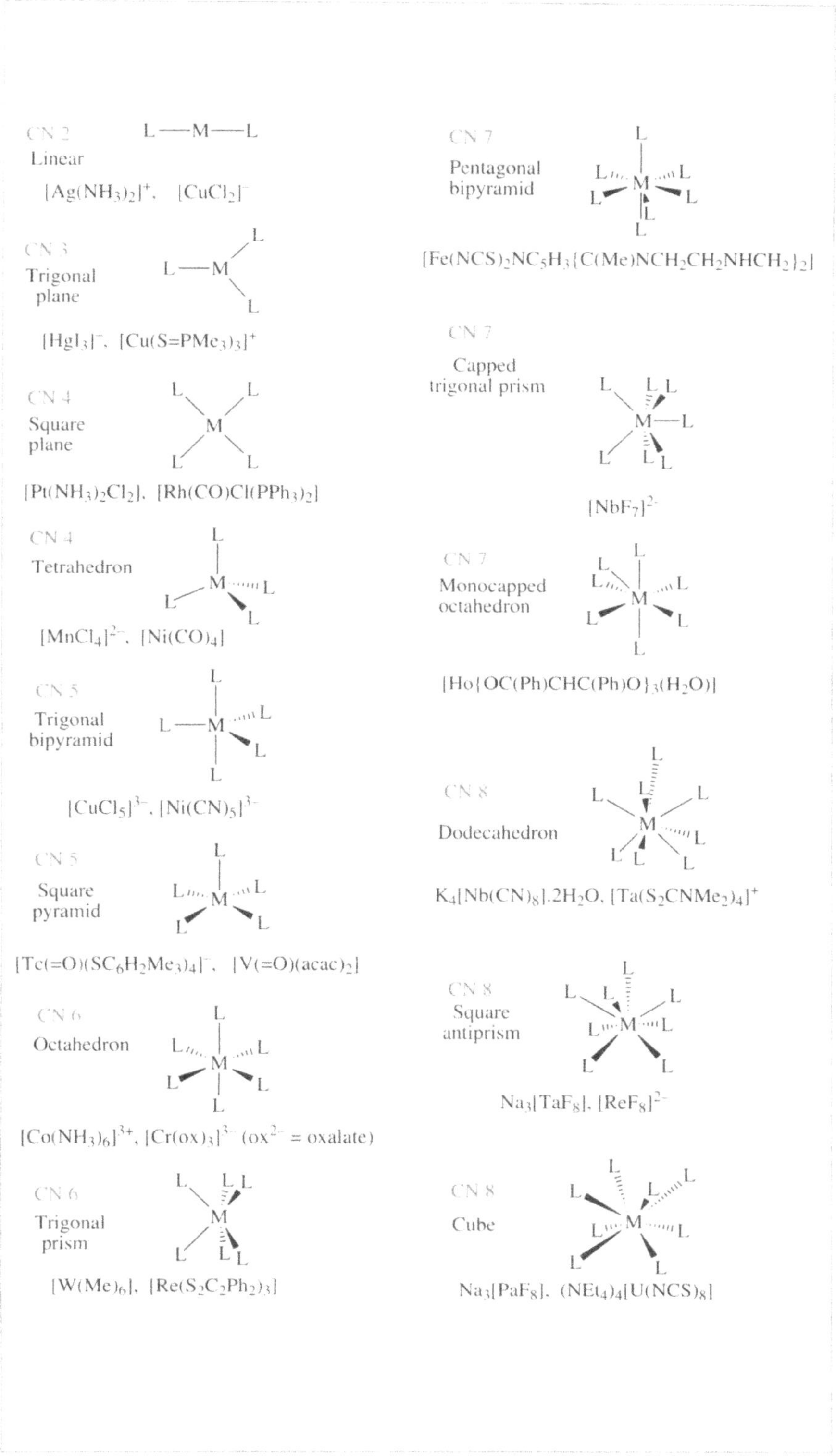

CN 2
Linear
L—M—L
$[Ag(NH_3)_2]^+$, $[CuCl_2]^-$

CN 3
Trigonal plane
$[HgI_3]^-$, $[Cu(S{=}PMe_3)_3]^+$

CN 4
Square plane
$[Pt(NH_3)_2Cl_2]$, $[Rh(CO)Cl(PPh_3)_2]$

CN 4
Tetrahedron
$[MnCl_4]^{2-}$, $[Ni(CO)_4]$

CN 5
Trigonal bipyramid
$[CuCl_5]^{3-}$, $[Ni(CN)_5]^{3-}$

CN 5
Square pyramid
$[Tc(=O)(SC_6H_2Me_3)_4]^-$, $[V(=O)(acac)_2]$

CN 6
Octahedron
$[Co(NH_3)_6]^{3+}$, $[Cr(ox)_3]^{3-}$ (ox^{2-} = oxalate)

CN 6
Trigonal prism
$[W(Me)_6]$, $[Re(S_2C_2Ph_2)_3]$

CN 7
Pentagonal bipyramid
$[Fe(NCS)_2NC_5H_3\{C(Me)NCH_2CH_2NHCH_2\}_2]$

CN 7
Capped trigonal prism
$[NbF_7]^{2-}$

CN 7
Monocapped octahedron
$[Ho\{OC(Ph)CHC(Ph)O\}_3(H_2O)]$

CN 8
Dodecahedron
$K_4[Nb(CN)_8].2H_2O$, $[Ta(S_2CNMe_2)_4]^+$

CN 8
Square antiprism
$Na_3[TaF_8]$, $[ReF_8]^{2-}$

CN 8
Cube
$Na_3[PaF_8]$, $(NEt_4)_4[U(NCS)_8]$

Figure A.1 Some examples of metal complexes for various coordination numbers

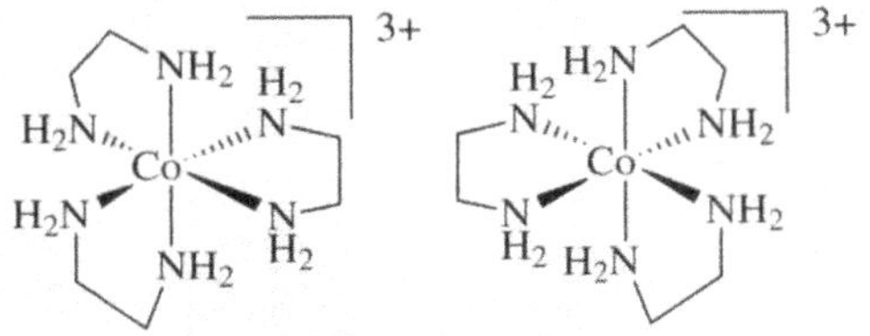

Enantiomers
A.2

4.3. (i) The possibilities here are tetrahedral or square planar, in which case *cis* or *trans* isomers are possible (Figure A.2a)

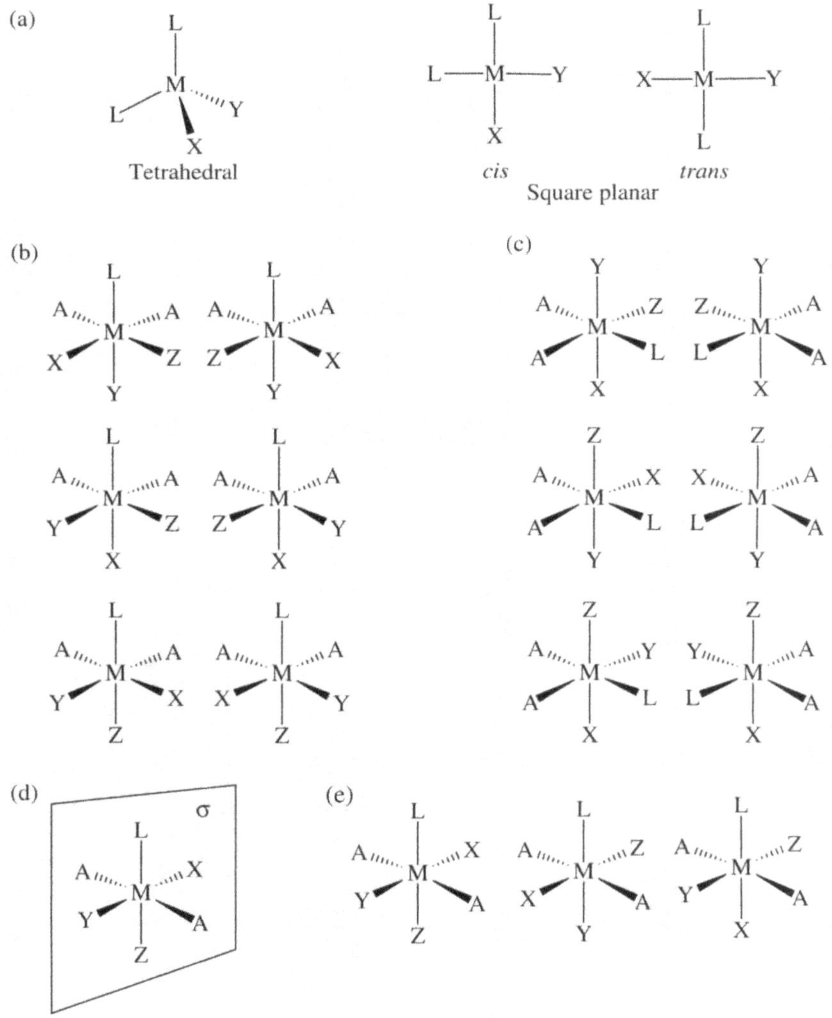

Figure A.2 Isomeric structures for (a) [ML_2XY], (b) *cis-fac*-[MA_2LXYZ], (c) *cis-mer*-[MA_2LXYZ], (d) the mirror plane in *trans*-[MA_2LXYZ] and (e) *trans*-[MA_2LXYZ]

(ii) $[MA_2LXYZ]$ is very unlikely to be trigonal prismatic, so only the octahedral case need be considered. The complex could have the LA_2 ligands in a *fac* arrangement with the two A ligands in *cis* positions, in which case the molecule is chiral with no symmetry elements, and so there are six isomers, in fact three pairs of enantiomers (Figure A.2b). Alternatively, the LA_2 ligands could be *mer* with the two A ligands *cis*, leading to another three pairs of enantiomers (Figure A.2c). Finally, the LA_2 ligands could be *mer* and the two A ligands *trans*, in which case there is a mirror plane (Figure A.2d) leading to only three isomers (Figure A.2e).

4.4. The possible isomers are shown in Figure A.3. In fact, two isomers of this complex have been isolated and studied by X-ray diffraction methods. Both have a slightly distorted square pyramid structure, one being orange (**A.3**), the other violet (**A.4**).

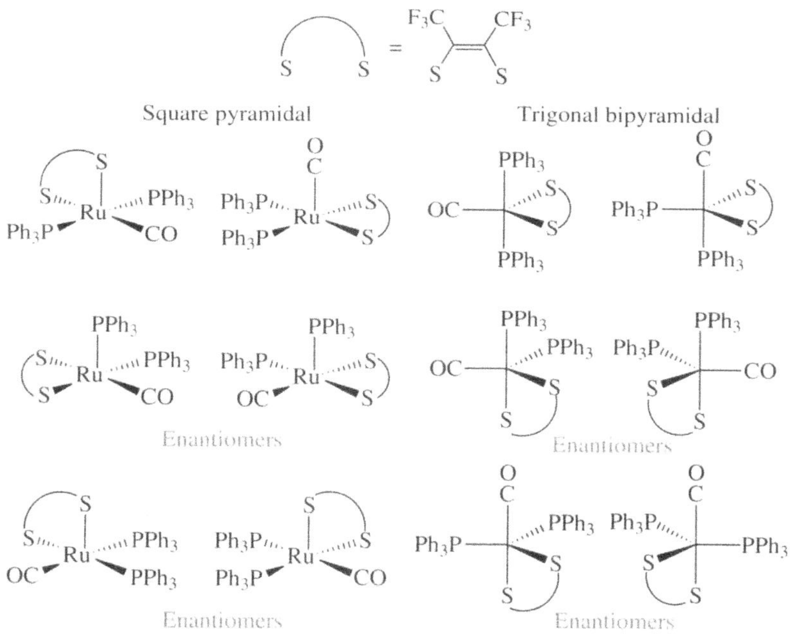

Square pyramidal

Trigonal bipyramidal

Enantiomers

Enantiomers

Enantiomers

Enantiomers

A.3

A.4

Figure A.3 Possible isomers of $[Ru\{S_2C_2(CF_3)_2\}(CO)(PPh_3)_2]$

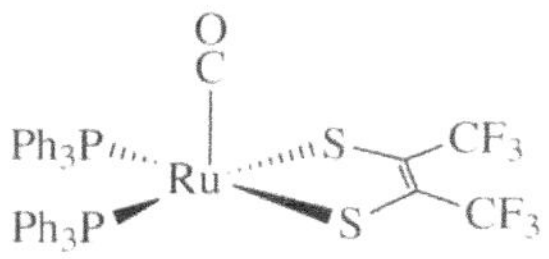

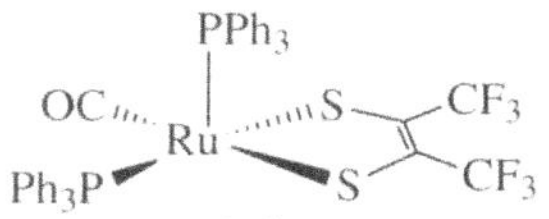

4.5. Compound **4.8** is *cis*-dichlorobis(triethylphosphine)platinum(II); compound **4.9** is *trans*-tetramminebis(dimethylamido)-cobalt(III) ethoxide; compound **4.10** is bis[*cis*-dibromo(μ_2-dimethyl sulfide)platinum(II)]; compound **4.11** is benzenethiolatochloro(η^2-propene)palladium(II).

Chapter 5

5.1. Equation 5.27 for the reaction of $\{AgF\}_{(aq)}$ with $\{Ca(NH_3)\}^{2+}_{(aq)}$ can be obtained by subtracting equation 5.28 from equation 5.29. Since $\log_{10}K$ values are proportional to free energies, they are additive, so $\log_{10}K$ for equation 5.27 can be obtained following the same logic, and is given by 2.96 – (–0.7) = 3.66, so that $K = 4.57 \times 10^3$ for equation 5.27.

5.2. (i) *Soft* metal ions form their strongest complexes with larger, more polarizable ligand donor atoms such as sulfur or phosphorus. *Hard* metal ions are those which form their strongest complexes with smaller electronegative donor atoms such as oxygen. See Table 5.1.
(ii) This just requires a qualitative answer based on a little reading. 'Soft' donors such as S or P tend to stabilize lower metal oxidation states, so the stability of Fe^{2+} compared to Fe^{3+} would be greater in an $\{N_5S\}$ coordination environment than in an $\{N_6\}$ coordination environment. This will facilitate the Fe^{3+} to Fe^{2+} reduction, moving the reduction process to more a positive potential, other things being equal. A model is provided by two synthetic iron porphyrin complexes, for which changing an axial histidine nitrogen donor atom to a methionine sulfur donor atom results in a positive shift in the Fe^{3+}/Fe^{2+} potential of 170 mV.

5.3. To solve this problem it is probably simplest to set up a cycle relating the various $E^\ominus$ and β values as shown in Scheme A.2. The free energy change for each side of the cycle can be obtained, either from $E^\ominus$ using equation 5.3, or from β using equation 5.1. Starting from Fe^{3+}, either way round the cycle to $[Fe(CN)_6]^{4-}$ involves the same free energy change, so that:

$$\Delta G^\ominus_{ox} + \Delta G^\ominus_{CN} = \Delta G^\ominus_{Fe} + \Delta G^\ominus_{rd}$$

Rearranging to get an expression for $\Delta G^\ominus_{rd}$ in terms of the other free energies and bringing in $E^\ominus$ or β values as appropriate gives:

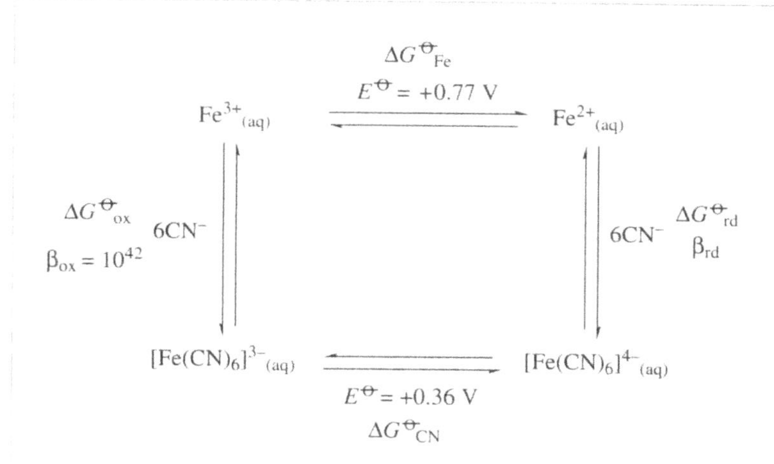

Scheme A.2

$$\Delta G^\ominus_{rd} = \Delta G^\ominus_{ox} + \Delta G^\ominus_{CN} - \Delta G^\ominus_{Fe}$$
$$\Delta G^\ominus_{rd} = (2.303RT)\log_{10}\beta_{ox} + \{-nF(0.36)\} - \{-nF(0.77)\}$$
$$= -2.303RT\log_{10}\beta_{rd}$$
$$\text{so } \log_{10}\beta_{rd} = 31 + (0.36)(nF/2.303RT) - (0.77)(nF/2.303RT)$$
$$n = 1 \text{ so that } \log_{10}\beta_{rd} = 31 + (0.36/0.059) - (0.77/0.059) = 24.05$$

Thus $\log_{10}\beta_{rd} = 24.05$ and $\beta_{rd} \approx 1.12 \times 10^{24}$ (to three significant figures).

5.4. Subtract equation 5.36 from equation 5.35 to obtain equation 5.34. Since $E^\ominus$ values are not additive, they need to be converted to their respective free energies for the two equations 5.35 and 5.36, $\Delta G^\ominus_{35}$ and $\Delta G^\ominus_{36}$, using equation 5.3. The free energy for equation 5.34, $\Delta G^\ominus_{34}$, can then be calculated. This can be converted to $\log_{10}\beta$ using equation 5.1:

$$\Delta G^\ominus_{34} = \Delta G^\ominus_{35} - \Delta G^\ominus_{36} = -nF(1.69) - \{-nF(-0.6)\} = -nF(2.29)$$

Thus for equation 5.34: $-2.303RT\log_{10}\beta = -nF(2.29)$

$$\log_{10}\beta = nF(2.29)/2.303RT = 2.29/(0.059) = 38.8$$
$$\beta = 6.31 \times 10^{38} \text{ (to three significant figures).}$$

5.5. (i) The *macrocyclic effect* refers to the finding that complexes of macrocyclic ligands are more stable than those of their acyclic counterparts. An example is provided by the greater formation constant of $[Ni(cyclam)]^{2+}$ compared to that of $[Ni(2,3,2\text{-tet})]^{2+}$ (Figure 5.4).

(ii) Equation 5.37 for the formation of $\{Ni(cyclam)\}^{2+}_{(aq)}$ can be obtained by adding equations 5.38 and 5.39. Adding the $\Delta H^\ominus$ and $\Delta S^\ominus$ values for equations 5.38 and 5.39 gives values for equation

5.37. $\Delta G^{\ominus}$ can be then calculated and $\log_{10}K$ evaluated:

$$\Delta H^{\ominus} = -71 + (-59) = -130 \text{ kJ mol}^{-1}$$
$$\Delta S^{\ominus} = 55 + (-47) = +8 \text{ J K}^{-1} \text{ mol}^{-1}$$
$$\Delta G^{\ominus} = \Delta H^{\ominus} - T\Delta S^{\ominus} = -RT\ln K = -2.303RT\log_{10}K$$
$$= -130 - 298(8)/1000 = -132.4 \text{ kJ mol}^{-1}$$
$$\log_{10}K = -\Delta G^{\ominus}/2.303RT = 132.4/5.706 = 23.2$$
$$K = 1.60 \times 10^{23} \text{ (to three significant figures).}$$

The formation of $\{Ni(cyclam)\}^{2+}{}_{(aq)}$ from $Ni^{2+}{}_{(aq)}$ is primarily driven by the enthalpy term for equation 5.37 with only a small contribution from the entropy term, although this also favours complex formation. A positive $\Delta S^{\ominus}$ term arises for the complexation of $Ni^{2+}{}_{(aq)}$ by 2,3,2-tet ($-T\Delta S^{\ominus} = -16.39$ kJ mol^{-1}), but the replacement of this ligand by cyclam is opposed by entropy with a negative $\Delta S^{\ominus}$ term ($-T\Delta S^{\ominus} = +14.0$ kJ mol^{-1}), possibly because 2,3,2-tet is more strongly solvated than cyclam and ties up more water molecules. The macrocyclic effect in this case is enthalpy driven, and reflects the 'good fit' of the Ni^{2+} ion in the macrocyclic cavity of cyclam. The entropy term opposes the reaction of equation 5.39.

5.6. This question primarily relates to charge/radius ratio effects on metal ion–ligand binding. Across the lanthanide series, $\log_{10}K_1$ values increase steadily as the ionic radii of the Ln^{3+} ions decrease. However, errors in the measurements mean that the correlation is not perfect. The smaller, more polarizing ions form the stronger complexes. The value for Y^{3+} suggests a radius close to that of Gd^{3+}, a little larger than structural studies indicate. A similar trend is seen for the An^{4+} ions, but $\log_{10}K_1$ values are larger than for the Ln^{3+} series because of the higher charge. In the cases of UO_2^{2+} and NpO_2^{+} the lower charges give rise to smaller $\log_{10}K_1$ values, although the metals are in higher oxidation states. However, with the actinides, and especially the higher oxidation state ions, a covalent contribution to bonding is possible. This being so, $\log_{10}K_1$ values may be higher than expected from simple charge/radius ratio arguments. In the lanthanide series, the core-like character of the 4f orbitals does not allow significant covalent contributions to bonding.

5.7. The term *equilibrium template reaction* is defined in Box 5.4, and an example is provided in Scheme 5.3. The term *kinetic template reaction* is defined in the margin note on page 93 and an example is provided in Scheme 5.4.

Chapter 6

6.1. In a trigonal bipyramidal complex with the x and y axes in the equatorial plane, it can be seen from Figure A.4 that the degenerate d_{xy} and d_{xz} orbitals are directed between the ligands and so will be lower in energy. The degenerate d_{xy} and $d_{x^2-y^2}$ orbitals lie in the equatorial plane defined by three of the ligands and will be at higher energy. The d_{z^2} orbital points directly at the axial ligands and the toroidal part of the orbital lies in the equatorial plane, so this orbital is at the highest energy.

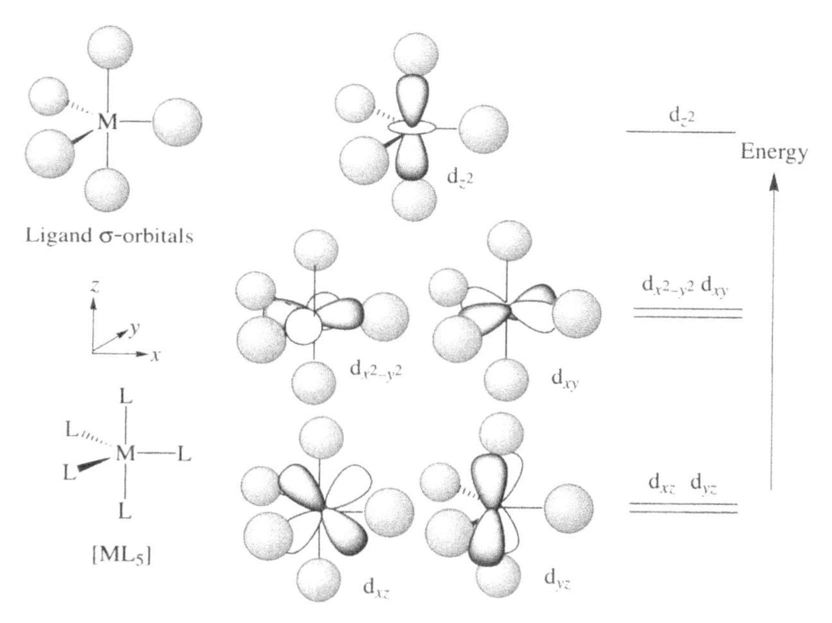

Figure A.4 The development of a crystal field splitting diagram for a trigonal bipyramidal complex, $[ML_5]$

6.2. (i) Chose five from the following list:

Geometry of the complex	Nature of the metal ion
Charge on the metal ion	Whether first-, second- or third-row metal
Nature of ligand	Charge on ligand
σ Basicity of ligand	π Basicity of ligand
σ Acidity of ligand	π Acidity of ligand

(ii) In each case this involves comparing $10Dq$ with the pairing energy, PE, to determine whether the complex is high or low spin. The appropriate crystal field splitting diagram should then be drawn to show the appropriate electron configuration. The value of the CFSE can then be calculated, remembering to take into account the number of *additional* pairing energies required in low-spin cases and converting units from cm^{-1} to $kJ\ mol^{-1}$, as shown in Scheme A3.

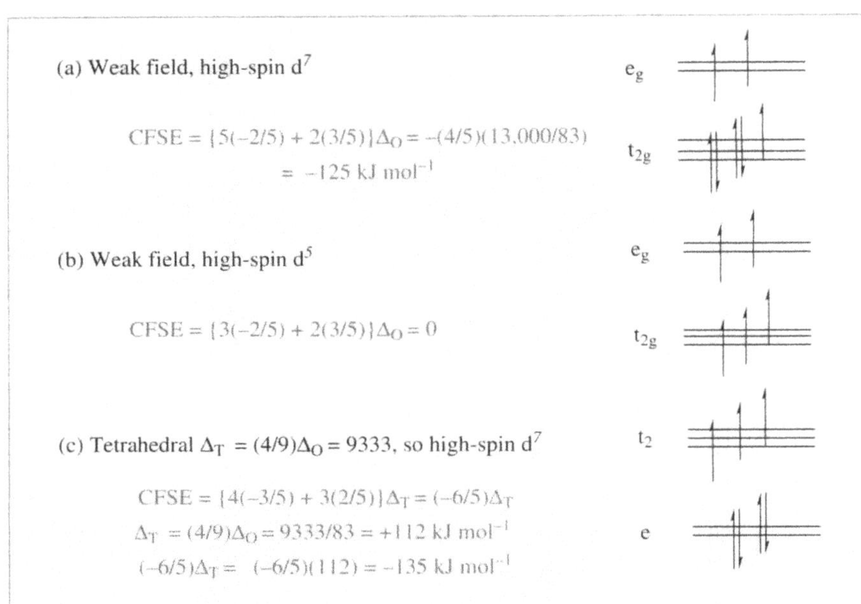

Scheme A.3

6.3. Since CN^- is a strong-field ligand, with d^6 Fe^{2+} it gives a $t_{2g}^6e_g^0$ configuration having no unpaired electrons, and the maximum CFSE. However, with d^8 Ni^{2+}, CN^- gives a $t_{2g}^6e_g^2$ configuration. Inspection of Figure 6.9 shows that, with a strong-field ligand, an octahedral structure is less energetically favourable for d^8 ions than a square planar structure, so $[Ni(CN)_4]^{2-}$ is formed and not $[Ni(CN)_6]^{4-}$.

6.4. The longer axial bond distances are an example of the Jahn–Teller effect. To explain this it is necessary to quote the Jahn–Teller theorem (see margin note on page 106), and develop the crystal field model for an octahedral metal complex (Figure 6.2), before showing the effect of a tetragonal distortion on the d orbital degeneracies and the energy of the system (Figure 6.7). In the case of a d^9 ion such as Cu^{2+} there is an energy benefit from elongating two *trans* bonds in a six-coordinate complex.

6.5. The starting complex is low-spin Co^{3+} d^6, so the ligands are in effect strong field. An octahedral crystal field splitting diagram (Figure 6.2) is needed to show this. A two-electron reduction produces a Co^+ d^8 system which, with strong-field ligands, will prefer a square planar geometry. The reasons for this need to be explained through reference to Figure 6.9. The non-chelating ligands Cl^- and py are the most likely to dissociate, leaving $[Co(dmgH)_2]^-$ as the square planar d^8 product complex.

6.6. A discussion of π bonding in octahedral complexes is required here with reference to Figure 6.16:

In (i), F^- is a good π donor and reduces Δ_O because the low-energy t_{2g} orbitals are occupied by 'fluoride electrons', leaving the higher-energy t_{2g}* orbitals to accommodate 'metal electrons'. In contrast, NH_3 is neither a π donor nor a π acceptor to any significant extent, giving a Δ_O value little affected by π bonding effects.
In (ii), CN^- is a good π acceptor and increases Δ_O because the low-energy t_{2g} orbitals are unoccupied and can accommodate the 'metal electrons'. In contrast, H_2O is a weak π donor, giving a Δ_O affected slightly by π bonding effects and in the opposite sense.

6.7. (i) Synergic bonding involves metal-to-ligand π back-donation of charge to relieve the charge build-up on the metal which results from ligand-to-metal σ donation, as explained in Section 6.3.2 and Figure 6.18a.
(ii) In order of increasing ν_{max}(CO): **6.3** < **6.4** < **6.5**.
Compound **6.3** is a neutral complex with the CO opposite a potential π donor ligand, and so has the largest back donation to CO π* orbitals, and so the lowest value for ν_{max}(CO). Compound **6.4** is a cationic complex with the CO opposite a potential π donor ligand, and so has intermediate back donation to CO π* orbitals, and so an intermediate value for ν_{max}(CO). Compound **6.5** is a cationic complex with CO opposite another π acceptor CO, and so has the weakest back bonding and the highest value for ν_{max}(CO).

Chapter 7

7.1. The four processes are intra-ligand transitions, ligand-to-metal charge transfer, metal-to-ligand charge transfer and d–d transitions. Refer to Section 7.2.1 for the explanation of these terms.

7.2. This weak absorption is due to the d–d transition within the d^1 Ti^{3+} ion. This is associated with the process $t_{2g}{}^0e_g{}^{*1} \leftarrow t_{2g}{}^1e_g{}^{*0}$ in terms of the occupancy of orbitals from the crystal field model, and represented in spectroscopic terms by the transition $^2E_g \leftarrow {}^2T_{2g}$. The energy of this transition corresponds with $10Dq$ or Δ_O for $[TiL_6]^{3+}$, and 19,200 cm^{-1} is 19,200/83 = +231 kJ mol^{-1}. An Fe^{3+} complex with the same Δ_O value and a pairing energy of +260 kJ mol^{-1} would be high spin with the electron configuration $t_{2g}{}^3e_g{}^{*2}$. This contains five unpaired electrons, so

$$\mu_S = 2\sqrt{\tfrac{5}{2}\left(\tfrac{5}{2}+1\right)} = 5.92\ \text{BM}$$

7.3. The weak band is a d–d transition, $^2E_g \leftarrow {}^2T_{2g}$ for d^1 Ti^{3+}, and the shoulder arises from Jahn–Teller splitting of the ^2E_g level. The more intense bands are charge transfer bands. Since X is a π donor, MLCT bands are not possible, so the other bands must be due to LMCT. There are four possible LMCT transitions: ligand $\pi(t_{1u})$ to metal t_{2g} and e_g* and ligand $\pi(t_{2u})$ to t_{2g} and e_g* (Figure 6.17). The energy order of the ligand π orbitals is not known but since $10Dq$ = 19,000 cm^{-1} from the d–d transition, the bands at 8000 and 27,000 cm^{-1} are associated with transitions from the higher-energy ligand π orbitals and the bands at 12,000 and 31,000 cm^{-1} with transitions from the lower-energy ligand π orbitals.

7.4. The solution to this problem involves deciding possible geometries, determining numbers of unpaired electrons from the crystal field splitting diagram, and calculating spin-only magnetic moments using equation 7.2.

(i) $[FeX_4]^{3+}$ contains Fe^{3+}, d^5; a tetrahedral structure may be assumed as square planar structures require strong-field d^8 situations. The electron configuration $e^2t_2^{\,3}$ gives five unpaired electrons, so

$$\mu_S = 2\sqrt{\tfrac{5}{2}\left(\tfrac{5}{2}+1\right)} = 5.92 \text{ BM}$$

(ii) $[FeZ_6]^{2+}$ contains Fe^{2+}, d^6, octahedral, low spin. The electron configuration $t_{2g}^{\,6}e_g^{\,0}$ gives zero unpaired electrons, so $\mu_S = 0$.

(iii) $[NiZ_4]^{2+}$ contains Ni^{2+}, d^8, a strong-field case. This gives a square planar structure with the electron configuration $(d_{xz},d_{yz})^4(d_{z^2})^2(d_{xy})^2(d_{x^2-y^2})^0$. Hence zero unpaired electrons and μ_S = 0.

(iv) $[NiX_6]^{2+}$ contains Ni^{2+}, d^8, a weak-field case. This gives an octahedral structure, $t_{2g}^{\,6}e_g^{\,2}$ with two unpaired electrons, so μ_S = 2.83 BM.

(v) $[CuX_6]^{2+}$ contains Cu^{2+}, d^9. Jahn–Teller distortion of the six-coordinate octahedral complex gives a tetragonal structure. The electron configuration is $(d_{xz},d_{yz})^4(d_{xy})^2(d_{z^2})^2(d_{x^2-y^2})^1$ with one unpaired electron, so μ_S = 1.73 BM.

7.5. (i) Since PE > $10Dq$ in $[Fe(H_2O)_6]^{2+}$, the octahedral d^6 Fe^{2+} ion must be high spin:

CSFE = $\{-(4 \times \tfrac{2}{5}) + (2 \times \tfrac{3}{5})\}\Delta_O = -\tfrac{2}{5} \times (14{,}000/83) = -67.5$ kJ mol^{-1}. With four unpaired electrons the complex has μ_S = 4.90 BM.

(ii) Since PE < $10Dq$ in $[Fe(CN)_6]^{3-}$, the octahedral d^5 Fe^{3+} ion must be low spin:

CSFE = $-(5 \times \frac{2}{5})\Delta_O$ + 2PE = $-\frac{10}{5} \times (24{,}000/83) + 2 \times (16{,}000/83)$ = -192.8 kJ mol^{-1}. With one unpaired electron the complex has μ_S = 1.73 BM.

(iii) Since $[MnCl_4]^{2-}$ is not d^8, the expected structure is tetrahedral and the d^5 Mn^{2+} ion must be high spin as PE > $\Delta_T = \frac{4}{9}\Delta_O$ = 8444 cm^{-1}.

CSFE = $\{-(2 \times \frac{3}{5}) + (3 \times \frac{2}{5})\}\Delta_T$ = 0 kJ mol^{-1}. With five unpaired electrons the complex has μ_S = 5.92 BM.

(iv) Since Co^{2+} in $[CoCl_4]^{2-}$ is not d^8, the expected structure is tetrahedral and the d^7 Co^{2+} ion must be high spin:

CSFE = $\{-(4 \times \frac{3}{5}) + (3 \times \frac{2}{5})\}\Delta_T$ = $-6/5 \times (\frac{4}{9})\Delta_O$ = -109.2 kJ mol^{-1}. With three unpaired electrons the complex has μ_S = 3.87 BM.

7.6. The Dy^{3+} ion has a $4f^9$ electron configuration, and the ground state term symbol $^6H_{15/2}$ shows that $S = \frac{5}{2}$, $L = 5$ and $J = \frac{15}{2}$. The value of g_J for Dy^{3+} can be obtained using equation 7.4 and the value of μ_J can then be obtained using equation 7.3:

$$g_J = \frac{3}{2} + \frac{\frac{5}{2}\left(\frac{5}{2}+1\right) - 5(5+1)}{2\left\{\frac{15}{2}\left(\frac{15}{2}+1\right)\right\}} = \frac{3}{2} - \frac{21.25}{127.5} = 1.3333$$

Therefore $g_J\sqrt{\frac{15}{2}(\frac{15}{2}+1)}$ = 10.6 BM to three significant figures.

Subject Index

Printed in the USA
CPSIA information can be obtained
at www.ICGtesting.com
LVHW081008230124
769544LV00045B/1009